The intersection of politics and spirituality in addressing the climate crisis

A dialogue with
Mohammed Sofiane Mesbahi

ALSO BY THE AUTHOR

Heralding Article 25: A people's strategy for world transformation

First edition

Copyright © 2020 Mohammed Sofiane Mesbahi

The moral right of the author has been asserted.

Apart from any fair dealing for the purposes of research or private study, or criticism or review, as permitted under the Copyright, Designs and Patents Act 1988, this publication may only be reproduced, stored or transmitted, in any form or by any means, with the prior permission in writing of the publishers, or in the case of reprographic reproduction in accordance with the terms of licences issued by the Copyright Licensing Agency. Enquiries concerning reproduction outside those terms should be sent to the publishers.

Matador
9 Priory Business Park,
Wistow Road, Kibworth Beauchamp,
Leicestershire. LE8 0RX
Tel: 0116 279 2299
Email: books@troubador.co.uk
Web: www.troubador.co.uk/matador
Twitter: @matadorbooks

ISBN 978 183859 281 3 (paperback)
978 183859 282 0 (hardback)

British Library Cataloguing in Publication Data.
A catalogue record for this book is available from the British Library.

Printed and bound in Great Britain by 4edge Limited
Typeset in 11pt Adobe Garamond Pro by Troubador Publishing Ltd, Leicester, UK

Matador is an imprint of Troubador Publishing Ltd

Dedicated to our true priests—the many activists fighting tirelessly to defend the rights of Mother Earth, God's holy temple.

"If enough people really did perceive the true extent of the climate crisis and its inner man-made causes, there would not be just a handful of committed activists among billions of other people. The minds of many concerned citizens may be engaged with this issue, their goodwill may be engaged, but you need to awaken the spiritual heart centre of humanity as a whole—which means you need the active and directed love of millions upon millions of ordinary people. You need the masses standing beside you, hearts aflame."

—Mohammed Sofiane Mesbahi

Contents

Editor's preface	xi
Part 1: A just transition through 'fair shares'	1
Part II: The inner and outer CO_2	27
Part III: Demonstrating love-in-action	65
Endnotes	94
About the author	95

Editor's preface

The following discussions with Mohammed Sofiane Mesbahi were initially conducted at the time of the Paris climate conference in December, 2015—the first truly universal deal to tackle climate change. Four years on, the views he put forth are, in many ways, more pertinent than ever.

While the Paris Agreement commitments are about to come into effect post-2020, the climate crisis has dramatically escalated in public awareness. At the same time, grassroots activism has suddenly grabbed mainstream news headlines, surprisingly led by a school-striking youth. It's notable that Mesbahi anticipated the need for such popular mobilisations in this interview, calling as he does for ordinary citizens to protest through massive and continuous demonstrations that press governments to shift towards zero-carbon economies in response to a planetary emergency.

Yet the impasse in the UN climate talks hasn't fundamentally altered since 2015. For example, civil society groups continue to uphold the principle of equity as 'the gateway to climate ambition', but wealthier countries are still manoeuvring to evade their fair shares of the necessary global action. Big corporate polluters still obstruct progress at the talks, and the world continues to have an overriding reliance on fossil fuels, which shows no sign of abating in the years to come. All this makes a mockery of the scientific consensus that global emissions must be cut in half over the next decade in order to keep temperatures within the safe limit of 1.5°C—still we remain on course for a catastrophic 3°C rise by the end of the century.

Although these core issues are outlined in Part 1, it must be stressed that the books real import goes beyond this initial policy-related discussion. In fact, it took many years to convince Mesbahi to be interviewed about STWR's broad position on international climate politics. This was in large part due to his longstanding and deep aversion to being interviewed in any form. But it is also because he considers the spiritual counterpart to our understanding of the broader environmental crisis—later described as the 'inner CO_2'—to be of far greater significance. The reader is therefore encouraged to embark upon a very different line of enquiry, one that may lead us to an expanded awareness about the deeper reasons why our modern societies have yet to transition towards more equitable and sustainable modes of living.

Altogether, this wide-ranging dialogue acts as an introduction to Mesbahi's pioneering vision of mass civic engagement towards a united cause for sharing the world's resources. The perennial themes raised throughout form the basis of further writings that are being published as part of an ongoing series entitled *Studies on the Principle of Sharing*, of which more can be found at www.sharing.org.

London, UK, November 2019

PART I

A just transition through 'fair shares'

What is your basic verdict on the Paris Agreement—was it justified to call it a 'major leap for mankind' and 'the world's greatest diplomatic success'?

There was some justification to the acclamatory news headlines, given the fact that an international climate treaty was finally sealed following the dramatic failure to agree a universal agreement in Copenhagen six years ago. The new language adopted around the goal to keep global temperatures at no more than 1.5°C above pre-industrial levels is also a testament to more than two decades of advocacy efforts and meticulous science. In the light of 21 years of stalled and often factious negotiations, I think few could disagree that this is a surprisingly ambitious target, one that has even taken the scientific community by surprise.

There were also some grounds to hail the talks as a success now that targets have been set around each nation's Intended Nationally Determined Contributions (INDCs), which are the first emission reduction plans under the UN Climate Change Convention (UNFCCC) that apply to both developed and developing countries, forming the official basis of a post-2020 global framework. From the most optimistic evaluation, the fact that we now have a long term goal of achieving net zero greenhouse gas emissions by the second half of the century is at least a timeframe for action, albeit with lots of concerns from civil society groups about the uptake of negative emissions technologies. The fact that developing countries made the most ambitious commitments, including China and India, also means there are signs that the geopolitics of global leadership has started to shift, and the South is willing to ramp up their mitigation efforts despite the unwillingness of the North to sufficiently 'take the lead'.

Whether the treaty will bring us any closer to a just and sustainable world order is another matter. Despite the aspirational 1.5°C emissions target, there is no clear roadmap for how to deliver these collective reductions in the short term. Even if the current INDCs are met by 2030, various studies have shown that we will still be on track for a 3 to 4°C warmer planet, leading us to extremely dangerous tipping points. The introductory text to the treaty itself admits this, stating that far greater emissions reductions efforts will be required if we are to address the significant gap between nations mitigation pledges and aggregate emissions consistent with even the 2°C pathway. There is nothing to prevent nations from reneging on their already insufficient commitments, which in the history of multilateral UN talks gives us little reason for optimism.

As expected, the only binding element of the agreement is for each nation to submit regularly updated goals on progress. It may be a legally binding document of international law as part of the UNFCCC, but there is no longer a legal responsibility for rich countries to provide finance to help poor countries adapt to climate change, let alone any legally binding targets for meaningful carbon cuts. So in reality, it is a sad indictment of our times that the Paris Agreement has been hailed as 'ambitious' and 'politically historic'. In this respect, we can also observe a striking parallel with the Sustainable Development Goals agreed in September 2015, which promises an 'end to poverty everywhere' by 2030 among other ambitious environmental targets, but without any sanctioning mechanisms or credible means to implement those outcomes.

From your standpoint at Share The World's Resources, it is interesting to note that the concept of 'fair shares' has now become a rallying call for action in the climate change debates. What do you see as the relevance of the principle of sharing to the Paris Agreement and the Conference of the Parties (COP) meetings in general?

The idea of sharing is emerging as a key theme in many areas of policy thinking and activism today, none more so than in civil society advocacy work on climate change. This is perhaps natural and to be expected, as the principles of fairness and equity are of course recognised in the UN Climate Convention, in that Annex I countries are expected to take the lead in emission reductions while respecting the rights of Annex II countries to sustainable development, which includes the right of less developed countries to receive financial and technical support. How to share responsibility for keeping global carbon emissions within scientifically-accepted limits, and on a fair and equitable basis, has always been at the heart of the COP process. The landmark equity principle of Common but Differentiated Responsibilities and Respective Capabilities (CBDR-RC) enshrines this recognition, based on the essential premise that all states are responsible for addressing global warming, yet not *equally* responsible, thus highlighting the profound moral issues that have caused so much contestation and division since negotiations began in 1992.

But that principle has now been all but abandoned in the Paris Agreement, which is a significant step backwards. Major industrialised countries waged a successful campaign against the equity principle by opting to determine their own targets on a purely national basis through the INDCs, and without any reference to the scale of global effort needed to stay within the 1.5°C target. This effectively erases their historical responsibility and militates against effective action from all countries in years to come. One of the key goals of the US in particular was to weaken the language in the text around 'loss and damage', so that they would not be liable for mandatory compensation for climate impacts in poor countries.

As for climate finance to help developing countries with adaptation and mitigation, it was already clear that the goal of raising $100 billion per year is woefully inadequate, and only a fraction of these sums have so far been delivered. But the new language in the text around climate finance being 'voluntary' and 'shared among all countries' further shifts the burden of responsibility back onto poorer countries. Relatively little finance is expected to come from guaranteed public funds, and there is the added risk that finance will be diverted from existing aid flows. The bulk of future financing may have to come from new market mechanisms such as carbon offsets which also risks an overreliance on private investments, as there is scant hope that governments will mobilise additional sources of revenue that flow directly into UN climate funds, such as from financial transaction taxes or a progressive carbon tax system.

Many of these points have been expanded on in much greater detail by civil society observers, but what should be obvious is that the COP21 outcome in no way reflects the principles of sharing and justice in a true form. Only the activists and campaigners in the conference halls seemed to embrace a vision of what equity means in addressing the climate crisis through urgent short-term actions and transformative, systemic change. Interestingly, civil society thinking on 'fair shares' has moved on significantly in recent years, based on a global carbon budget that accounts for historical responsibility and emissions allocations for individual countries. But Northern governments have now moved even further away from accepting the need for a new paradigm for equitably sharing the world's shrinking atmospheric space, and as such the prospect of enshrining a high-ambition carbon budget into a legally binding climate treaty is, alas, still a dream.

Can you expand on the relevance of a carbon budget in formulating an equitable solution to preventing dangerous levels of global warming?

Despite the present failures of multilateralism and the inadequacies of the UNFCCC negotiations, I believe it remains imperative that a vision for a fair climate deal is upheld by campaigning groups and progressive analysts, in accordance with the latest science. The question of a global carbon budget is central to this vision, as it reveals how much carbon can be emitted into the atmosphere without breaching the internationally-agreed upper limits on global warming. Hence if the remaining budget is divided on the basis of fairness and equity, it poses a number of pressing questions about how to share the remaining atmospheric space among the world's nations, particularly with respect to the differing levels of development between the global North and South.

The carbon budget approach has been discussed for many years among NGOs and academics, and is now well established in science following the IPCC's fifth assessment report, which released a budget assessment for different temperature limits based on data up to 2011. The UN estimated that about 1 trillion tonnes of carbon can be emitted into the atmosphere if we are to have a decent prospect of remaining within the

2°C limit, which reduced to about 800 billion tonnes when accounting for additional warming factors or other greenhouse gases. Of this amount, the IPCC calculated that we've already used up more than 500 billion tonnes due to carbon emissions from human activity since the onset of the industrial revolution, which means that at least half of the budget is already spent. On the current emissions path, the remaining carbon space was predicted to be exhausted within 2 to 3 decades.

However, various studies have updated these assessments and calculated that we have far fewer years remaining to stay within the 1.5°C budget. It is well accepted within both the mainstream scientific and NGO community that 2°C does not represent an adequately safe threshold to prevent 'dangerous' climate change, yet most of the scenarios work to date has focused on 2°C or 3°C limits, and the science is less robust on the safer 1.5°C limit. There could be less than five years remaining of CO2 emissions at current levels before we exceed the budget for 1.5°C, and even that gives us a mere 66% probability of avoiding the risks of tipping points and severe impacts on food and human security. It's notable that the 66% probability considered a 'likely' chance in the IPCC terminology is highly questionable, and if we apply a risk level of the kind that would be deemed acceptable in other areas of human activity, say 90% or higher, then there is actually no carbon budget remaining for keeping below 1.5°C.

It's also worth emphasising that the IPCC estimates are extremely complex due to the different assumptions and methodologies used, and there are a lot of uncertainties about the feedback processes that amplify or reduce the warming we see in the atmosphere, hence the need for various probabilities of staying below a certain temperature, none of which can be guaranteed. New studies are often released that find the IPCC estimates are too generous, and that we may be overestimating the total carbon budget by anything up to 200%.

The political, economic and social implications are enormous when you look at how to apply a scientifically-defined and internationally-agreed cap on total global emissions, and how to then distribute the remaining carbon budget between nations or among the population. It's already clear from the official UN data that the challenge of limiting emissions to even the 2°C threshold is extraordinary. An equitable sharing of the mitigation and financial efforts towards these ends will require major sacrifices from the Annex I nations, a renewed focus on the critical social and economic needs of developing nations, and a degree of international cooperation that is without precedent in human history. But as I mentioned, we still appear to be far from acknowledging the true extent of this great civilisational challenge. Political leaders in Paris failed even to discuss a total carbon budget as a basis of targets and effort sharing, while fossil fuels companies are still being facilitated in investing massive sums into the development of new reserves that will make it impossible to keep temperatures within a safe budget.

Please do comment further on what a truly equitable and ambitious level of international cooperation should look like on a country-by-country basis. What do you see as the specific achievements of current civil society thinking on how to translate the principles of sharing, justice and equity into a multilateral climate regime?

A lot of attention has now been given to this question among civil society groups, which has evolved from the 'climate debt' debates in activist circles that came to prominence around the time of the Copenhagen summit in 2009. The question is how to develop an ambitious climate regime that is "in the light of equity and the best available science", as long accepted as a principle in the UN Climate Change Convention and reaffirmed by world leaders in the Paris Agreement, but not yet translated into a quantitative global framework. A number of NGOs have therefore looked at how to operationalise the core equity principles of the Convention, fundamental to which are the concepts of historical emissions and historical responsibility.

One of the commonalities of the various proposals discussed is the concern for economic justice in any approach to climate protection, which gives rise to the need for an equity-based effort-sharing framework that safeguards the right to sustainable development for countries in the global South. The politics of the COP negotiations have proven that this is not merely an ethical priority, but the very basis of geopolitical realism and the gateway for increased environmental ambition—a point that has been well argued by many civil society observers.

In the highly regarded Greenhouse Development Rights (GDRs) framework, for example, the right to sustainable development is codified by way of a 'development threshold', meaning a level of income per capita that should not be accounted for in calculations about a country's capacity to tackle climate change. In other words, a country's capacity can be defined as the sum of all individual incomes, excluding incomes below the development threshold, which should be higher than the official global poverty line so that it can reasonably apply to all citizens of both the North and South, and somewhat reflect an adequate standard of living. The persuasive reasoning is that people below this level of income, those who have yet to realise their right to development, should not have to bear the burden of a climate transition.

To further account for the inequitable historical emissions between the North and South, the GDRs framework also proposes the use of equity indicators to calculate each nation's responsibility, as well as its aggregate capacity. For example, historical responsibility is calculated by the cumulative per capita greenhouse gas emissions of each country since an agreed start date, such as 1850 or 1990, which is adjusted to account for the development threshold. In this way, a country's fair share of the global mitigation effort can be defined by combining its responsibility and capacity, thus generating a single indicator of obligation. The real dynamic potential of such an index of course depends on first defining the remaining carbon budget, for which a 1.5°C marker pathway should clearly be the basis of negotiations.

My brief explanation here of the GDRs methodology is rather cursory and incomplete, and I would recommend referring to their literature to better understand how the use of equity indicators can provide a quantifiable, fair system for sharing the global effort among all countries. But I think their key achievement is to demonstrate how a major commitment to North-South cooperation is essential for any viable climate change mitigation framework. The Annex I countries have such a comparatively larger share of global responsibility and capacity that they cannot possibly meet their fair share of effort through domestic action alone. Thus they are duly obligated to provide less developed countries with the finance, access to technology and capacity building that is necessary for them to exploit their full mitigation potential, in line with their sustainable development strategies.

In the latest iteration of such an equity-based framework prior to COP21, a civil society review of the INDCs was able to illustrate just how far the pledged actions of wealthier countries compare to their respective fair shares of effort, central to which is the need for vastly scaled-up means of implementation. The INDCs of the US and EU, for example, only represent about a fifth of their fair shares. But it should be emphasised that even these estimates are conservative and pragmatic, based as they are on a 2°C global mitigation pathway and with reference to the wholly inadequate INDCs, leading civil society organisations to recommend a 'ratcheting-up mechanism' that can enable deeper, and possibly legally-binding commitments to be made in the future.

Then again, we cannot expect developing countries to accept a binding mitigation framework unless the principles of sharing and equity are at its heart. It must be seen that poverty eradication and human development can go hand in hand with a Marshall-style transition to a zero-carbon energy supply, as informed by the scientific reality. There is no way around the impasse: Northern countries will need to take a greater share of the burden at first, and face up to their obligations for massive international transfers of technological and financial support to poorer countries in the South. What we are talking about, really, is a new vision of international cooperation that is somewhat akin to the Brandt Report of 1980.[1] It is high time that report was updated to reflect the reality of a world that is fast exceeding ecological boundaries, in which a global reallocation of resources is needed to address both the climate and poverty crises simultaneously.

The Brandt Report proposed a kind of Marshall Plan for the Third World based on global Keynesian-style stimulus measures, but is it possible for the remaining carbon budget to be shared on the basis of equity, thus achieving the 1.5°C or even 2°C target, without a significant contraction of the global economy in the longer term?

This is a question that is not given much credence in mainstream political and academic circles, where the primacy of economic growth is seldom contested. But I agree with the basic assessment of many green economists and environmental think tanks, like the Club of Rome and the UK's former Sustainable Development Commission, that the continued ramping up of global economic activity is at odds with our attempts to avoid dangerous climate change and sustain the planet's ecosystems. There is also the Tyndall Centre research institute in the UK which has addressed this issue from a carbon budget analysis, making a convincing case around the need for immediate and planned 'de-growth' strategies of reduced consumption and economic contraction in the Annex I countries.

Their analysis is worth reflecting on, as it demonstrates how meeting the 2°C commitments will require Annex I countries to decarbonise their economies drastically and with immediate effect, not further delaying action until, say, 2020 when the INDC pledges of the Paris Agreement begin. It is only on this basis that developing countries may be enabled to peak in emissions by 2025, and yet still minimally grow their economies while embarking on a rapid transition away from fossil-fuelled development. According to the cumulative emissions budget approach that the Tyndall Centre adopts on the basis of the equity principle, wealthy nations must reduce emissions by 8-10% per annum over coming decades, which is at a rate far below what is considered by most economists to be compatible with a growing economy. There is, in fact, no historical record of any country succeeding in prolonged emissions reductions of even 4% each year without experiencing a deep economic contraction. The Stern report by the UK's Committee on Climate Change memorably observed that annual reductions greater than 1% are historically associated with economic recession or upheaval.

Hence the conclusion from the respected analysts at the Tyndall Centre that a planned period of austerity and rapid decarbonisation is necessary in the US, EU and other wealthy nations, in order to compensate for continued economic growth and increasing emissions in the poorer nations. There is no longer the time for a gradual, evolutionary transition to low-carbon energy supplies, which could take another 2 or 3 decades to completely put in place. So we must deliberately seek an equitable reduction of energy and resource consumption, while adopting more directly redistributive economic strategies in place of the conventional pursuit of economic growth.

This presents a challenge, no doubt, to the failed assumptions of *laissez-faire* globalisation, and it points to the need for a paradigm shift in economic orthodoxy and a new theory of macroeconomics beyond the obsession with GDP measurement, all of which is well discussed in the radical fringes of public debate. I would also add that the Tyndall Centre's analysis is again very conservative in its assumptions, based as it is on a 2°C and not a 1.5°C pathway, and only accounting for a 50% probability which is exceedingly low, as I remarked upon earlier. If we also account for the growth in emissions over the several years since they developed their scenario pathways, then the decarbonisation rates for wealthy nations may need to be significantly higher than even 10% per annum, perhaps in the range of 15% to 20%. The implications are sobering, to say the least.

What are some of these broader implications if continued economic growth worldwide is not viable due to environmental constraints alone, notwithstanding resource constraints and other ecological limits? If you believe in the creation of a world where everyone gets their fair share of the earth's resources, does this not imply a relative convergence in living standards between and within all countries?

From an economists standpoint, one of the major implications of the research and analysis we've been discussing is that the assumption of most policymakers that carbon intensity can be reduced sufficiently to allow for a continued increase in production, or so-called 'green' or 'sustainable' growth, is highly refutable. The belief that we can decouple greenhouse gas emissions from economic growth in the longer term is by no means held up by the evidence, in which permanent and absolute decoupling has been shown to be rare, if not fictitious. Innumerable studies cite the phenomenon of the rebound effect, whereby efficiency gains tend to be reinvested in more growth and consumption, hence negating the purported benefits in decreasing emissions.

Furthermore, if we consider the fact that wealthy nations effectively export their emissions and production to other countries, then it paints a different picture entirely. Once you take full account of all consumptive-based emissions, including from international aviation and shipping, then the level of decoupling for most countries is almost insignificant, especially if set against a rapidly shrinking global carbon budget. This altogether points to a highly contentious conclusion, but I think it's obvious to presume that we cannot continue to expand the world economy year on year, or universalise the current levels of affluence that are expected in wealthier industrialised nations, and yet still meet the carbon commitments set out in Paris. Whatever productivity gains are achieved through energy efficiency technologies and renewables, absolute reductions in emissions of the scale required also depends on a dramatic reduction in global energy consumption. And by implication, that depends on governments adopting new macro-economic policy goals that are no longer predicated on continuous GDP growth.

A lot of theoretical and modelling work has now been done around the concept of a steady state economy, but as you point out in the question, the real issue is how to achieve and manage the transition to equitable global sustainability. It's not only a question of how to fairly share the global emissions budget, or how to distribute the rights to pollute the atmosphere; there are also far larger, and even more difficult considerations around how the natural world should be used and shared. How can we create a world in which everyone gets their fair share of resources while we are already pushing against several key planetary boundaries, and the world population is rapidly growing? Ecological footprint analysis has made some interesting findings in this field, demonstrating how our combined demand for resources already exceeds the bio-capacity of the planet by 50 percent. We also know that it is the richest 20 percent of consumers who are appropriating the vast majority of global resources, and contributing by far the most to environmental degradation.

The ideal of the 'fair earth-share' may be rather quixotic in this regard, but it brings to life a very inconvenient truth—

that high-income countries may have to reduce their per capita ecological footprints by up to 80 percent, if anything close to a convergence in material living standards is to be achieved across the world without breaching environmental limits. Right now, of course, this visionary concept of a converging world is mainly the preserve of high-minded social scientists, considering that the world continues to diverge in terms of most indicators of wealth and inequality.

It appears that as an international community we do not want to face up to the immense implications of achieving a balanced distribution of world resources within the biophysical reality, particularly with regards to our national governance systems that were never built to manage trans-boundary problems on the basis of genuine cooperation and sharing. I would say it's clear that both the climate and broader ecological crises are compelling nations to rethink the entire model of ever-expanding global trade based on endless economic growth and our high impact, energy intensive and consumer-driven lifestyles. But the real question concerns how to actually initiate a voluntary transition towards more equal, participatory and ecologically resilient societies, regardless of all the thinking that's been done on the policies and tools needed for this enormous societal shift.

Let's return to this question in more depth later, and for now taking the discussion back to the Paris climate talks, what do you see as the main obstacles to achieving the kind of fair and ambitious multilateral regime that you outlined earlier?

The main obstacle is surely self-evident for any campaigner within the climate justice movement, in that the COP negotiations are dominated by powerful vested interests, while rich governments are politically captured by the fossil fuel lobby and corporate class. In December last year the Paris summit boasted an unprecedented level of corporate sponsorship, which was reflected within the conference halls by the renewed focus on market-based and technology-driven solutions like biomass energy carbon capture and storage. Carbon markets are now back on the table in a big way it seems, suggesting that governments still believe they can trade their way out of the climate crisis instead of committing to a drastic reduction in emissions in the short-term. The real solutions were never going to be part of the agreement, like immediately ending fossil fuel subsidies, transitioning to diversified agroecological food systems, relocalising our economies and planning a global shift to 100 percent renewable energies.

I think the contradiction at the heart of the Paris Agreement is apparent to most observers, in that the final document employs a lot of nice language around 'deep reductions in global emissions' or 'sustainable patterns of consumption and production', yet it basically ignores the systemic political and economic roots of the environmental crisis. So while governments were negotiating a climate accord and pledging to accelerate the reduction of their country's emissions, they were also continuing to negotiate secretive and environmentally harmful trade deals, like the Transatlantic Trade and Investment Partnership (TTIP). According to leaked documents, the European Union blocked all discussion in the Paris talks of any measures that may restrict international trade, and as we know the many plurilateral agreements being negotiated behind closed doors are geared towards a significant increase in fossil fuel imports and exports.

There is also a clear contradiction in that the UN climate process focusses only on the demand side of fossil fuel consumption, but pays no attention to its production. So it was unsurprising that the Paris Agreement said nothing about the immense spending on fossil fuel subsidies each year, which is now in the realm of $5 trillion. Nor is it a surprise that the whole idea of keeping fossil fuels in the ground was not even acknowledged in the final agreement, despite the huge amount of campaigning and public support around this issue. Pollution from international transportation is also entirely excluded from the agreement, which is inconceivable in light of how much the shipping and aviation industries are expected to contribute to global emissions in coming decades.

For the ordinary person of goodwill who tries to make sense of these byzantine climate negotiations, the one obvious conclusion should be that we cannot trust our heads of state whose agenda is overshadowed by what I describe as the forces of commercialisation. Just as we don't know what happens behind the closed doors of dark and devious trade deals like the TTIP, no-one on the street knows what is really discussed amid the corporate circus of the COP process. I often call these government leaders our politico-accountants, for as soon as they go home they will immediately sign more contracts for major polluting corporations whose interests they invariably uphold.

To take our own government in the UK as an example, they signed up to the Paris Agreement with much fanfare, and on the very next day they announced sweeping cuts to renewable energy subsidies, and committed a U-turn by opening up new areas of the country to fracking, even under national parks and wildlife protection zones. President Obama also signed off an end to a 40-year ban on crude oil exports just days after the Paris climate accord was agreed, which was a massive giveaway to the oil industry that will further exacerbate carbon emissions. So who can deny that these politico-accountants in our governments are completely overshadowed by the ideology of unbridled market forces, and hence they cannot be blithely trusted to act on our behalf with an issue as important as saving the planet?

In the face of these entrenched corporate forces and powerful vested interests, what can ordinary people do to try and steer government policy in the right direction?

This is the time when activists and engaged citizens must again, again and again organise massive demonstrations and direct actions to compel governments to shift towards zero carbon economies as an overriding priority. We need the public to come in as a whole and embrace the necessity of rapid social transformation, almost like a huge boycott of consumerism, of the very idea of business-as-usual. The inspiring protest activities we saw in Paris during the climate talks, regardless of President Holland's opportunistic ban on peaceful demonstrations, is exactly what we need to see—although on a far, far larger scale in every city.

We should now be protesting continuously around the world, not just once or twice each year or only during the Conference of the Parties. Total cumulative emissions have risen at an unparalleled rate in the new millennium, and as just discussed it is naïve or foolish to believe that world leaders will take effective mitigation action without immense pressure from the global public. Let's also not forget what happened after the international stock market crash in 2008, when governments immediately put environmental issues on the shelf in order to bail out collapsing banks and resurrect the economy. What do we think will happen if there's another serious global financial crisis, as many leading analysts are predicting?

If we look holistically at the scale of environmental problems we face, it is clear that huge and constant demonstrations around the world are the only solution for reordering governmental priorities. It's not that we as the public are dependent on our governments—actually the opposite is true, as governments are dependent on the present lack of public engagement so that they can continue to prioritise major corporate and financial interests. From another perspective, world leaders have actually played their part by coming together and agreeing the headline goal of a 1.5°C temperature limit; now it's up to the public to continue demonstrating every night and day until commensurate action is taken to achieve the necessary social, political, economic and technological transformations.

It's also notable that every big demonstration about the environment not only makes a noise that is heard by governments, but they help to inform the general public and create a wider awareness of the environmental emergency. It has a powerful effect. The media has an important role to play in educating the public in this regard, and if they were to constantly talk about the need for social transformation to avert a climate catastrophe, then we may see demonstrations around the world becoming monumental in the face of government inaction. Of course the mainstream media outlets rarely act in this way, with few exceptions—such as the ground-breaking action by newspapers around the world to speak with a common voice prior to the Copenhagen conference in 2009, many printing a front page editorial that called upon governments to share the burden of fighting climate change on a more equitable basis.[2]

In general, however, the mainstream media has failed to fulfil its vital role by educating people about the very real extent and urgency of addressing the climate crisis. On the contrary, most of the media are too busy spreading false propaganda on behalf of their corporate and political benefactors, if they don't ignore the issue entirely. Again in the UK, for example, none of the tabloid newspapers—those read by the majority of the public—led with a front page story about the 'historic' climate deal in Paris on the day after it was agreed. Yet this is an issue that may determine the future survival of humanity, requiring fundamental systemic and lifestyle changes that will eventually require the cooperation and participation of everyone.

Over recent months and years, tens of thousands of people have participated in a global wave of actions to keep fossil fuels in the ground, with a concerted call for a just transition to a clean energy era. Do you see this resurgence of rallies, protests and civil disobedience as a sign that the public is waking up to the reality of the climate crisis, or is there a long way to go before we see the kind of broad-based social movements that you are envisioning?

Indeed on one hand, we have a climate deal that isn't truly ambitious enough. But on the other hand, we have a global public that isn't ambitious enough. For example, in September 2014 around 400,000 people gathered in the streets of New York City calling for bold action from governments to address the climate crisis, which was the largest climate march in US history. But there are more than 300 million people in America. So you might say that while 400,000 people were busy protesting, at the same time there were hundreds of millions of people who were busy consuming. The same can be said of the dozens of other countries where big climate marches have recently taken place, which remain relatively tiny in proportion to the population as a whole. In truth, the greatest danger for the environment is not only the polluting corporations or their beholden governments; it's also the complacency or sheer indifference of the public at large.

To be sure, there are many committed activists who campaign with maturity and intelligence about the need to keep fossil fuels in the ground, to switch to decentralised and renewable energy sources, to develop a more localised and ecologically-resilient economy, to protect the commons from corporate enclosures, and so on. But they are a comparatively small number of people trying to do a job that requires the backing of the whole population. They're effectively on their own trying to do the job for everyone else, which doesn't make any sense when you listen to the warnings from scientists who say we're heading for a climate catastrophe.

Those very few activists are doing what they can to save the world—the rest are best described as consumer-citizens who leave the world's problems to other people and their governments. So we may observe that the government is overshadowed by the forces of commercialisation, while the consumer-citizen is overshadowed by their own complacency and indifference. Within this tiresome and depressing reality, it is understandable that many active people of goodwill feel jaded and overwhelmed in trying to fight for the greater good of all.

Environmental campaigners should be aware of another fact, however: it was in large part their activities and worldwide demonstrations that compelled world leaders to achieve a universal agreement in Paris, especially in light of the public outrage that arose after the failed climate deal in Copenhagen. Pope Francis's strong critique of inequality and overconsumption is also an important factor that has urged world leaders to acknowledge the people's voice. The Pope's counsel on the environment issue should be heeded very carefully by believers and agnostics alike, as reflected in his recent Encyclical letter which makes an effective case for sharing the world's resources in order to resolve the environmental crisis.

What the Pope essentially recognises is that we cannot combat our ecological problems unless we also combat the enormous discrepancies in living standards throughout the world, which calls for a sense of global solidarity and interdependency that is sadly lacking in human affairs. His encylical makes clear that the responsibility for transforming society lies not only with establishment politicians, who tend to be driven by powerful financial interests and generally lack a breadth of vision. The real responsibility lies with us all to overcome our social conditioning that sustains a profligate consumerist lifestyle, and instead to develop new attitudes to life that reject the materialistic mindset and the prevalent 'techno-economic paradigm'.[3] That can begin, I suggest, by joining with the millions of other people out there who stand for a simpler, saner and more equal world.

During the talks in Paris, many activists remarked on the immense gap between the urgent demands of civil society, and the reality of those 'politico-accountants', as you call them, who consistently put the interests of the rich and powerful before the interests of the most vulnerable, or indeed all of humanity. How do you see this gap lessening if we are going to successfully transition to a sustainable economy—are you saying that civil society groups must unite with a single voice in order to generate the political will needed for a complete transition to renewable energy?

The broad vision we outlined for a fair climate deal is never going to succeed so long as those politico-accountants are holding the reigns in governments, and even if they do eventually accept such commonsensical proposals it will take a very long period of time. And time is what we do not have if we need to maintain a 350 parts per million [ppm] greenhouse gas concentration in the atmosphere, when we've already exceeded an alarming 400 ppm. It's clear from climate scientists that the next decade is critical for taking decisive action to limit global warming, but it appears that we've reached an impasse when you look at that seemingly unbridgeable gap between political 'realism' and the need for rapid systemic transformation.

A lot of people believe the way forward is for social movements, environmental and development NGOs, and all progressive civil society organisations to join together in a combined front to challenge governments to take the necessary action, and we've seen a lot of initiatives and manifestos towards that end, all of which come and go and generally amount to very little. Unfortunately, those initiatives will never work unless a huge majority of the public is backing these groups through massive protests and peaceful direct actions that are attended by people in literally their hundreds of millions, which is currently far from the case.

It may even be detrimental to the work of many NGOs if they were to pour all their energies into the fight for government action on tackling climate change, as there is so much injustice and suffering throughout the world that the divergent causes of these many organisations are critically needed at this time. As I see it, their work is on the same line of energy as the governments and multinational corporations, where one is exacerbating the wrong trends towards further commercialisation, social division and environmental destruction, while the NGOs pull in the other direction and attempt to ameliorate the resultant harms. It is verily one fight, but a fight which takes many different directions, so much so that all the NGOs cannot harmonise their collective efforts when they are too busy fighting in their own ways, and for their own particular causes.

So it is wrong thinking to believe that NGOs and civil society groups must collectively lead the way in the present context of widespread public apathy and inertia. It is even a complacent way for the ordinary citizen to look at the problem, when what we need is for the public as a whole to realise the gravity of the situation of global injustice and environmental degradation. Then the public will empower the NGOs until their collected voice becomes so loud that governments will be compelled to heed their advice, in which sense the backing of the public will symbolically represent the fact that NGOs are united in their diverse aims.

PART II

The inner and outer CO_2

How do you explain the mindset of those powerful leaders and corporate executives who continue to profit from the climate crisis or promote fossil fuel industrial interests, despite the growing consensus of opinion that the majority of fossil fuels must be left in the ground?

From the most holistic assessment, what we are really witnessing is a war going on between the old age and the new, as represented by diverging left-wing and right-wing politics, or the increasingly polarised perspectives among staunchly progressive and reactionary attitudes. We are also witnessing increasing social divisions and global conflicts, as brought about by extreme and worsening inequalities on both national and global levels. Hence from within this maelstrom of escalating crises and confusion, it is difficult for the average citizen to comprehend the consciousness of billionaire corporate executives, like the Koch brothers, who will stop at nothing to sustain their fossil fuel industrial empire through lobbying activities and climate change denial.

The one who is profiting from the earth's destruction is also never going to understand the consciousness of those who are fighting against them, like the activists who opposed the Keystone XL pipeline or those who advocate for fossil fuel divestment. We are coming to a time when no-one will be able to understand the mindset of those who benefit from today's climate crisis, which would include the politicians who profess to be committed to an ambitious climate deal while they continue to negotiate behind the public's back for new contracts with polluting corporations. So rather than trying to understand the intentions of those self-seeking individuals who represent the old ways of a dying era, we would do better to understand the thinking of the environmental activists who are fighting for our collective future, and then soon enough join them.

Considering how clear is the economic, environmental and social case for shifting society towards 100% renewable energy and less resource-intensive lifestyles, it may actually seem mysterious to some people why their governments fail to implement the necessary policies to urgently steer this great transformation of society. Although progressive thinkers can explain this lack of sufficient political action in terms of the restrictions imposed by the current economic system and its ideological dominance, there is something much deeper going on behind the scenes that cannot be reduced to theoretical analysis of economic globalisation or neoliberalism.

In esoteric and spiritual terms, the reality is that new forces of life are fast entering the world, and establishment politicians don't have the intelligence and grace to recognise that the old competitive, corrupt and power-thirsty ways of the past are now coming to an end. The most extreme proponents of those old ways talk of nothing but making their countries great once again, thus continuing to engage in the fight for power through 'isms' of the left-wing and right-wing, which ultimately holds back the expression of goodwill and cooperation among the nations. As always, it is the world's poor and the environment that becomes the collateral damage of the policies that stem from these ignorant attitudes.

Alas it is very rare to find wise men and women in mainstream politics today, those who are consciously or intuitively aware that it's time for politicians to embrace the new forces entering the world that call for economic and political transformation based on sharing, justice and right human relations. Ultimately, the causative factor at the root of climate change is a question of human consciousness or awareness, and this problem is intimately related to all of the interlocking crises that define our time, which are essentially spiritual in their nature.

It will be interesting to explore in more detail what you mean by saying that climate change is spiritual in its nature, and stems from the problem of human consciousness. Is there a value for progressive thinkers and engaged citizens in fusing a spiritual understanding of life with political, economic and environmental issues? This appears to be a major theme of your ongoing work, *Studies on the Principle of Sharing*, which examines the problems of humanity from more spiritual or esoteric perspectives.

Here you are asking me to take an entirely new direction in this interview, with a very different kind of energy and dynamic. However I will be glad to do so because my main concern is not to be a commentator on the policy details of the COP negotiations or other UN summits, but rather to help stir a global uprising of the public on the basis of a spiritual understanding of our global interconnectedness, which I often refer to in terms of the one Humanity. The intersection of politics and spirituality is little discussed or understood in most activist circles, even though an awareness of the inner side of life is imperative if we want to grasp the reality of how a just transition to a sustainable economic system can actually happen.

As we just discussed, the ideological proponents of the old political way of thinking, based on economic power, competition and the selfish acquisition of resources, are unlikely to be interested in teachings about how to transform the world through massed goodwill and a true education based on Self-knowledge. But we have reached a time when the intelligent advocates for a better world, and especially the youth, are being called upon to look at political issues with a more holistic, inclusive or spiritual faculty of perception that is rooted in an awareness of the heart and its attributes.

If we are sincere and interested to understand the profounder spiritual implications of sharing the world's resources, then we have to learn to adopt this new understanding of the heart in our advocacy work and political theorising. It doesn't mean that we have to give up our diverse causes for social and environmental justice, or else begin to read certain esoteric philosophies alongside our political pursuits. There are many spiritual teachings that may be worthwhile for us to read, but it is more a matter of enquiring for ourselves into the meaning of right human relationship within our divided and fatefully commercialised world, which can only benefit us in our current activities and add depth to our understanding about how society needs to change. So I am primarily concerned with how ordinary people of goodwill can take a different route in their thinking, with this more heart-engaged kind of energy and new awareness of the interdependency of humanity as a whole.

Regarding environmental issues, we can think of this as the inner and outer CO_2, or the two different but interrelated forms of pollution that threaten the future sustainability of our planet. The outer pollution is interminably discussed in terms of the atmosphere and the natural environment, but the bigger problem is the inner pollution that has determined the activities of mankind throughout the millennia, culminating in our present-day civilisational or spiritual crisis. In other words, the greed, selfishness, gross ambition, arrogance, complacency, indifference, prejudice, hatred and so on—these are the internal attitudes and intentions that have shaped our collective destiny and continue to result in the tragic environmental effects we see all around us. The outer CO_2 doesn't happen by itself—it is clearly the consequence of the inner CO_2 that motivates our thinking and actions, from the individual up to the national and intergovernmental levels.

Hence it is ultimately impossible to bring about a safe and balanced environment by solely fighting against the outer activities of big corporations and governments, when the prevailing 'inner' values of our society remain largely beholden to materialistic and self-centred concerns. How can we limit the global temperature rise to below 1.5°C, when the drive for profit, power, wealth and luxury—as expressed through the intentions of countless millions of individuals—is already pushing us to overheat the planet to 4 degrees and even higher? Yet only the more mature environmental campaigners are preoccupied with this question of how to transform the inner attitudes of mankind, as is only just beginning to be reflected, for example, in civil society thinking about the importance of working with our cultural values to influence social change.

What are the concrete ways that we can benefit by trying to understand and act upon this awareness of the 'inner CO_2' and humanity's interdependency, particularly with respect to the global environmental crisis?

We were previously talking about the seeming mystery behind the reason why governments and major corporations aren't taking much more concerted action to address the climate crisis, such as by leaving fossil fuels in the ground and urgently transitioning to a world run on clean energy. And we reasoned that in order to really answer this question, we cannot only look towards the 'outer' explanations in terms of a systemic analysis of what is wrong with global capitalism, and how the economic system needs to be transformed; we also need to look at the inner or psychological causes of the problem and how man himself needs to change.

The value of pursuing this line of enquiry should be self-evident for anyone who is attuned to the deeper spiritual nature of world crises, and it demands that we learn to look at humanity's problems with a more holistic awareness that connects both our heart and mind in unison. Please bear in mind, however, that the heart and its attributes are never complicated but can only be understood on the most simple basis, which partly explains why an 'inner' awareness of world problems often fails to make an appeal to intellectuals and hardened political activists.

To adopt this more heart-engaged way of thinking in accordance with the above example, we can observe that a shift to 100 percent renewable energy and simpler living standards can be viably achieved within a relatively short time-frame, but it is not a question of whether or not the political will is there: more fundamentally, it is a question of how much love is there. Because to perceive all the human intentions that lie behind the outer activities around industrial fishing and agribusiness, for instance, or rainforest logging, strip mining and mountain top removal, it is clearly not motivated by an attitude of love or reverence for nature; it concerns only the mindset of money and profit through conforming to an economic system based on market forces and complex financial incentives.

This talk of 'love' is not to get lost in wishy-washy ideas about how we should all be nice to one another and live in peace with nature, as what we are really interested to perceive is the underlying psychological causes of social injustice and environmental degradation that originate from within our dysfunctional societies. 'Love' in the most basic spiritual and psychological sense means 'not to harm', whereas the very intention behind the drive for money and profit is harmful in all directions, both towards each other and the world around us, as well as towards ourselves.

If the company directors of a massive tar sands operation approach a beautiful and remote countryside, they would obviously not destroy that natural environment if they were motivated by an attitude of love and reverence for all that exists in this world, and they would rather seek to do something else with their lives that can help to heal, rather than exacerbate, the degradation of the environment. When analysed with an inner awareness of world problems, the only reason why these individuals can make huge profits from their activities is because they do not know themselves or what they have inside, meaning the true spiritual reality of the inner Self, for it is a lack of Self-knowledge and awareness that underlies all of our problems in the final analysis.

There may be a widespread appeal for many people in examining the world's crises from such a human and basically spiritual point of view, but can you be any clearer about the practical benefits for the progressive activist or scholar in adopting a more heart-engaged way of thinking in relation to their varied activities and causes? What is the real significance of this line of enquiry for those who yearn and fight for a better world?

To help you answer this question for yourself, look again very closely at the phrase 'political will' that you used earlier. Meditate on it awhile, and you will find that the phrase itself engenders a sense of hopelessness and even despair, for it is wrapped in the illusion that the present order of things is immutable and without an alternative. When we discussed how we are witnessing a war today between the old ways and the new, the entire etymology of the phrase 'political will' belongs to the old ways of the past Cold War period, and even earlier to the time of classical economists like John Locke and Adam Smith. It inherently means that change only happens from above, by the actions of our political leaders and those with power and influence, who with enough pressure from the public below may eventually make a small conciliatory gesture at the expense of elite interests.

Now contrast that phrase 'political will' with the idea of 'love-in-action', and try to envisage the significance of what this means with respect to the critical world situation. One is for the future, for the new forces and unifying energies that are fast entering the world, while the other represents the old order and institutions that are now rapidly breaking apart amidst the chaos of our interlocking systemic crises. Hence what we need most today is not the mere 'will' of our governments to implement the necessary policies, which is tantamount to asking those politico-accountants who currently run our nations to save the planet on our behalf. What we need is the manifestation of love and wisdom in the world, for if the energy of love was the prevalent motivating factor behind the collected actions of mankind, then there would be no such thing as the politico-accountant, nor the political activist for that matter. I have written before that the atheist cannot exist without the countervailing belief of others in God; and by a parallel analogy, the political activist cannot exist without the absence of love in our world.[4]

I am not suggesting that political activism is somehow futile or misguided in its present forms, which is quite the reverse in light of our escalating social and environmental turmoil. But if we want to perceive for ourselves what 'love' really means on a planet with plentiful resources in which millions of people are dying in poverty, in which the problem of man-made climate change is near to approaching a catastrophic climax, in which cold-blooded theft and hatred is rampant in international affairs—then we need only engage with our common sense and our inborn compassionate awareness. What we are observing is so banal that it is wise, for the truth of the heart is always unadorned and simple. Love in the context of world problems today is closely related to the need of detachment; meaning, in this respect, detachment from greed, from selfishness, from the old ways of aggressive competition, environmental destruction and legitimated theft.

Hence to understand the practical benefits of adopting this inner perspective and awareness, we should carefully ponder anew that seemingly mysterious question: what is preventing our societies from transitioning to clean energies and environmentally sustainable, more equitable living standards worldwide, in light of how feasibly that transition could be rapidly achieved? To repeat, the answer is not complicated in spiritual and psychological terms—it is simply and always a lack of love. For without the manifestation of love in this world, without the awakening of the spiritual heart centre of humanity as a whole, then it is impossible to propose a more enlightened social order that will be accepted and embraced by the populace in its majority. Any policies that are imposed from above to limit consumption levels or share resources will inevitably be resisted, opposed and ultimately rejected. And do we really imagine that such policies will be legislated for and enacted by our present governments?

So what this means for us personally could not be more pertinent, if we apply this line of enquiry to the question of worldwide social transformation. For example, when preparations for the Iraq war began in 2003 following the tragedy of 9-11, the entire world's attention was focused on that single event. Now what do we think it will take for the entire world to focus on the danger of runaway climate change with the same level of responsiveness and heartfelt concern, for surely the environment issue is even more foreboding in its implications than 9-11? In a similar vein, we should ponder what it will take for mainstream society to focus its unremitting attention on the reality of global hunger, which is a crisis that goes hand in hand with our environmental problems and could also be decidedly resolved, if only enough love was there. But considering the rampant complacency and indifference of humanity in which so many people don't want their comfortable 'way of life' to be disturbed, the greater significance of this inner awareness of world problems can only be left to our intuition.

There seems to be a growing number of individuals and groups that recognise the need for a more spiritual response to the problem of climate change, one that is based on a new meta-narrative of our interconnectedness and an attitude of stewardship and reverence of nature. Can you explain your view of the origins of the global ecological crisis from the inner or spiritual perspective that you're broadly defining?

For many centuries humanity has been living through its illusions, and now those illusions are fast breaking down the whole world appears to be confused, disorientated and without a clear way forwards. Many of the beliefs and creeds that sustain the old way of thinking are now being called into question through new scientific and anthropological findings, if not through sheer common sense. This would include the culturally engrained belief that greed and selfishness is the driving factor in human evolution, or that competition and inequality is the natural order of society, or that war and poverty is an unalterable fact of life. We are all familiar with the illusions that are sustained by these outmoded beliefs, even if we perceive their fallacy within our own life experience. Who would readily profess today, for example, that wealth is the key to happiness, or that power and success is the just reward for hard work, a competitive spirit and personal ambition?

Yet figuratively speaking, politicians and the business community act like a huge industrial unit that constantly manufactures these illusions, to the extent that they have developed an inadvertent systemic procedure for maintaining business-as-usual, which is to constantly invent new forms for the old illusions whenever the existing forms no longer work. It is instructive to observe in this light the focus on corporate technologies and market-based solutions in addressing climate change, as if we can solve the climate crisis by continuing to pursue the same profit-oriented, competitive and materialistic mindset that created the crisis in the first place. From the inner perspective, humanity is so conditioned by the old illusions that we refuse to realise that the world's crises—like the massive influx of refugees and migrants across Europe, the growing extremes of intra-country inequality, or indeed the environmental crisis in its totality—are forcing us to awaken to reality and completely change our whole attitude to life. But humanity appears to be so stubborn that we continue to seek money, power and success even as the world is erupting in chaos all around us.

Underlying all our illusions is the belief that man could divide himself from the environment, with his intention to have power 'over' nature by manipulating its laws and trying to possess its freely distributed abundance, rather than seeking to live simply and harmoniously within its self-regulating processes. But these are very petty powers that man has developed over nature, however impressive it may seem that we can excavate the earth for its riches and build enormous cities resplendent with luxury. Even our most sophisticated technologies are not developed by working 'within' the powers of nature, and hence they are all contributing to the slow devastation of the Earth by either direct or indirect means. Unless man completely changes his inner attitudes he is effectively digging up his own burial ground, for nature *per se* is not an illusion and will ultimately turn against us, or at least that is how it may seem.

Of course the deeper spiritual truth is that nature cannot seek vengeance against the human race for its collective transgressions, when humanity is forever integral to nature's balanced functioning. Man is Life, the integral midway point between the higher spiritual and lower kingdoms of nature—animal, vegetable and mineral. Thus when man tries to dominate the natural environment and overly exploit its limited resources, causing devastation and immense suffering in his wake, he is not only dividing himself from everything that exists in nature, but also from the spiritual reality of the One Life which he is an inextricable part of. So it is not the weather that needs to change, but only man himself—by finally perceiving the extent of his illusions and wrong intentions. Put another way, we have to learn how to know ourselves and leave things as they are, instead of trying to bend the natural world to our errant wills in the name of profit and grandiose delusions of human prowess and technological progress.

We are not only seeking power over the laws of nature, but bullying nature with our almost psychotic identification with illusions of man's supremacy and separation from the environment. Certain despotic individuals in history are well-known for their identification with hubristic illusions of personal power and greatness, although we now live in a world that is dominated by huge multinational corporations that embody similar illusions on an institutionalised and even grander scale. Hence these vast profit-seeking organisations are foreseeably more dangerous in their intentions, considering how they can influence laws and government policies to their own ends. We can also observe how this is the indirect purpose of corporate lobbying, marketing and advertising as an industry—namely, to create and sustain society's illusions however harmful the results for the natural environment and man's spiritual growth through Self-awareness.

To be clear, it is not our identification with ideas, beliefs or 'isms' that we are concerned with here, but the widespread identification in our culture with dangerous illusions that are holding back the spiritual evolution of humanity as a whole. We may have begun our quest for knowledge and power over nature with a kind of innocence since the Enlightenment period, and no doubt there are many talented scientists and technologists today who remain very sincere in their intentions to achieve dominion over the earth by unlocking nature's secrets. This mindset is also propagated through the entire edifice of our educational establishments, where even those who receive an elite schooling are conditioned to, in effect, despoil the planet and walk over other people in their quest to become rich, powerful, esteemed or professedly successful.

I have discussed elsewhere how those who receive a supposed 'good education' are in fact perpetuating today's disastrous social and environmental trends, particularly through being encouraged to achieve a high social status and thereby conform to the selfish arts of commercialisation.[5] Indeed, it is the visionless drive to create never-ending profits that has become the presiding factor in world affairs since the 1980s, leading to such divisions and destruction that we are finally forced to confront our inability to dominate the natural environment and pillage its resources without constraint. We could describe the combined forces of commercialisation as the purveyors of mankind's great illusion, deluding everyone from the ordinary man who dreams of owning a big mansion and fast cars as a route to happiness, to the corporate lobbyists in Capitol Hill and the Pentagon's war-planners who seek power and profit throughout the world, whatever the cost.

All of us are servants to these forces to some degree and sustain them by conforming to society's illusions, even as the environment is metaphorically on its knees, begging for mercy. And still a large segment of humanity is seemingly oblivious to the malicious effects of rampant commercialisation, which is a fact that we can only attribute to the formidable power of illusions. If these illusions continue to overwhelm our collective consciousness, then it will be like a giant super-computer that gradually takes over the minds of man, until we find ourselves witnessing the eventual onset of a nuclear world war. However far-fetched such a prospect may seem at the present time, please use your intuition to ponder the final destination where the illusions of commercialisation are leading us, then see if you can disagree with this observation.

It may be helpful to expand further on what you mean in saying that we all serve the forces of commercialisation to some degree, and therefore we are all part of this great illusion. In what way do we, as individuals, sustain the existing socioeconomic order of global capitalism, and how are we ourselves contributing to the eventual prospect of all-out warfare between nation-states, which may indeed seem implausible in this new millennium?

Many of us blame our governments, the corporations or 'the system' for world problems, without realising that we are an integral part of that system and sustain it through our own thinking and actions. Our education and conditioning, whether it's through schools or our social environment, is actually training us to be pliant consumers who perpetuate the system in its currently destructive form. How we all serve the forces of commercialisation should therefore be readily understood.

For example, by consuming products and services from a globalised marketplace regardless of its unjust structural arrangements, as most people do in Western society and within the affluent parts of developing countries, we are indirectly participating in a multifaceted system of exploitation and destruction. This is a fact that many people today acknowledge and understand, hence the whole movement of fair trade, consumer boycotts and so-called ethical consumerism. Surely most educated people are well aware that multinational corporations are responsible for grabbing land, clearing rainforests, poisoning rivers and polluting the environment, or exploiting the disposable labour force through low wages, poor working conditions, and so on.

We may blame our governments for supporting these practices through their various subsidies, tax breaks and other incentives, and we may also blame the big corporations for mesmerising us to consume all the myriads of products they sell for endless profits. But the fact remains that the vast majority of ordinary people are willing participants in the great illusion of rampant commercialisation. In our growing masses, we support and exacerbate the unjust arrangements of the global economy through our daily patterns of mass consumption, hence the illusion that this way of life can be sustained is perpetuated in a vicious cycle, repeating itself and worsening in its trends with each passing day.

However much we try and efface the deeper reality by placing an etiquette on top of our activities in terms of social norms or contemporary fashions, there is no escaping the fact that we are all somewhat responsible, however indirectly, for destroying nature and exploiting other people through our conformity with the existing consumer-driven economic system. The wanton feasting and consumption that surrounds our seasonal festivities are only the most pertinent examples of how most of us conform to the great illusion, whatever our political persuasions or activist causes.[6]

Can you perceive how there is a psychological equivalence between our heedless consumption habits and our votes for new government leaders, in the sense that we often blindly consume products without knowing or even caring about their origins, just as we vote for politicians and expect them to sort out the world's problems on our behalf? So we know that big supermarket chains are exploiting small farmers across the world, which means that every time I purchase the vegetables they produce, I am effectively making a vote for that exploitation. And even if I don't vote for an establishment politician that comes to office, I am still voting for them indirectly if I conform to the society they run without taking action to change their policies.

We like to call this system 'capitalism' and blame it for all of humanity's problems, but we have reached a time when we need to question what capitalism really means in this world of such tremendous violence and cruelty. Is all the suffering we witness around us today really the fault of capitalism *per se*, or is it the result of tremendous human-induced violence in all directions—towards the natural environment, towards the animal kingdom, towards each other and towards ourselves?

If we are able to perceive the world's problems in these more holistic and psychological terms, then we may accept that our present socioeconomic system based on rampant commercialisation does, in fact, constitute a war. Many analysts talk of the possibility of a third world war happening through escalating military conflicts, but in reality the gravest war on Planet Earth is taking place already as a result of the combined greed, self-centredness and indifference that is exacerbated by commercialisation. It is ultimately a war against life itself, against humanity's spiritual evolution, and yet almost everyone is participating in this slow-motion war to a greater or lesser extent. How, then, can we implement any blueprint or framework for a sustainable global economy, when mass consumerism and unrestrained commercialisation is the presiding factor in world affairs—and when we ourselves are the unwitting participants in this self-destructive violence?

Going back to your initial observations about the spiritual dimensions of the world's interlocking crises that are seldom discussed in political or activist circles, perhaps you can talk in more detail about what you described as the 'inner CO_2' that is the deeper origins of humanity's problems?

However much data and evidence we garner about ecological boundaries, atmospheric pollution and climate disruption, we have still not learnt the most basic lesson of how to live and evolve on this planet without disturbing the elements of nature. The ineluctable truth is that nature as a whole is a living being, as various scientists and philosophers have long hypothesised, although it is not just the material and objective world that man is polluting and destroying with his activities. There are two worlds, both visible and invisible, and it is the hidden side of nature that we are refusing to listen to or acknowledge as a reality, despite all the Ancient Wisdom teachings given to humanity down the ages. Just as we cannot see gas or CO_2 with the naked eye, there are untold invisible entities within nature that exist and have yet to be scientifically proven, although they play a vital role in regulating and sustaining our planetary biosphere in both its microcosmic and macrocosmic manifestation. Hence in our ignorance we continue to cause devastation to the environment by preventing these hidden elements of nature from carrying out their proper function, which is a fateful consequence of man's wrong turn in attempting to have power over his environment.

To give an important example, so many environmental scientists today are focused on the Earth's atmosphere and its interrelation to other natural systems, but few are concerned with the amount of noise that humanity is producing on a global level. Yet from the inner psychological or esoteric spiritual perspective that we are exploring, the incredible noise that is emitted by all countries of the world has an extremely deleterious effect on the hidden elements of nature, which in turn has a deleterious effect on the known elements of nature as revealed in the wind, rain, oceans and so forth. To perceive the reason why all this accumulated noise is so damaging to the environment, we need only ponder the effects of its origin—such as the activity of heavy manufacturing and extractive industries, the millions of automobiles, planes and container ships in perpetual transit, or the unrelenting conflict and destruction taking place within war-torn regions.

But it is not only the outer effects of that immensely noisy activity that damages the environment, for the accumulated noise that humanity produces also reflects how our societies are dysfunctional and disturbed, which in turn is directly related to the dysfunctional and disturbed state of the environment. To live in a world where there are so many wars, so much pain and suffering, so much arrogance and recklessness in our destruction towards each other and the sub-human kingdoms—all of this turmoil reflects itself upwards into the planet's biosphere and affects the hidden elements of nature. The well-known Hermetic principle states 'As above, so below', but today this axiom is better understood in reverse when contemplating the sorry state of the world; for as it is below with all the stress caused by man's divisive thinking and wrong intentions, so it is above. In other words, the inner side of man is as much a cause of our planet's environmental disequilibrium as man's outer activities within society. And the inner side of humanity is so dysfunctional that it almost resembles a mental and emotional sickness.

To give another example, you may recall how the people of Pakistan were jumping for joy on the day their nation created a nuclear bomb, with some people even crying in the streets, which is just a small indication of how human intelligence and emotion is going in the wrong way on our planet. What national pride can do, you might say, when the general public is ecstatic at the prospect of building a weapon that could kill many thousands of innocent civilians in another country. And in the midst of those celebrations the planet was experiencing devastating wars and conflicts, the pain of millions of people suffering in poverty, and all the stress and anxiety of human life in modern society—all of which is mirrored back to us through climate disasters and weather disruption.

Does this mean that the problems in human relationships, from the conflicts between individuals and within groups to the many inequalities and injustices that are experienced throughout society as a whole, are a direct cause of environmental disequilibrium—as much or even more than the damage we inflict on the environment through our industrial activities and mass consumptive behaviours?

Another esoteric truth is that every problem in the world has a corresponding sound or note, and today the frictional nature of those sounds are being reflected back to us through the unpredictability and turmoil of climate change and natural disasters. To understand this in simple terms, try to picture how the whole world is in conflict on every level, and ponder the effect that must have on the wider environment. If we imagine two people violently arguing in front of us, that atmosphere is also part of our inner climate on a psychological level, which is just as important to be aware of as the outer climate within the skies above us. And psychologically, socially, the inner climate that humanity moves within is in a disastrous and chaotic state, just as the outer conditions of the environment are now in a constant state of flux and chaos.

Taking this a step further in our awareness, we were previously observing how the great illusion of rampant commercialisation is masking the reality of tremendous violence, but it should be emphasised that such violence is psychological in its nature, as well as social and economic. So the person who is living under constant stress, who is depressed with having to work long hours in manual labour like a machine, is really the victim of a greater systemic violence that he experiences psychologically. And the effect of that all-pervasive violence has a profound impact on the elements of nature, because everything is interrelated and interdependent between man and his environment, from the inner to the outer.

So the selling of armaments between governments is clearly an act of violence, for what else can it be? But in the same way, the very existence of homelessness on our streets is not only a social injustice, but also an act of violence that is perpetrated indirectly by governments and the rest of society. If we look at this inner reality more deeply with an understanding of our psychological and spiritual interconnectedness, the desire of affluent nations to become more 'prosperous' when millions of people are hungry in the world, when millions more lack the basic essentials of life, that too is an act of violence. When the Western world celebrates Christmas through overconsumption and gluttony while so many people are dying from poverty in farther countries, that too is an act of indirect violence. Our collective complacency is an act of inconceivable violence towards the countless unknown people who are less fortunate than ourselves, notwithstanding its implications for the wider environment. We somehow accept all these forms of social, economic and psychological violence as business-as-usual, as a normal part of our everyday lives, and yet the accumulation of these inner attitudes and behaviours is what we see all around us. Climate change, extreme poverty, wars and conflicts—this is the eventual package of what you get when humanity ignores the need for right human relationship and lives in such a dysfunctional way over many decades, indeed for millennia.

We have now reached a point when the exhausting behaviours of humanity have caused such chaos and imbalance within our societies, that now the weather conditions on this planet resemble a mind that has become too stressed to function normally. Thus it may appear as if we have turned the environment into our enemy, forcing it to fight back through all the havoc created by tsunamis, earthquakes, floods, droughts and other natural disasters. And that havoc created by the environment is causing even greater havoc within our societies, which many analysts predict could lead to further wars between nations over diminishing natural resources like water. The behaviour of the environment reflects the behaviour of man, and what we now see happening is just the beginning. So unless man changes his ways completely by implementing the principle of sharing into world affairs, again we can foresee that the greatest danger for the future will not be caused by runaway climate change, but only by man himself through the unleashing of global warfare that may eventually destroy all life on earth.

You've just made an ominous prediction—that if humanity continues along the present course, it will not be climate change that eventually leads to a planetary catastrophe but rather our own actions as individuals and nations. Is there a further layer of tragic irony to this observation, in that many people today remain heedless of the great illusion of commercialisation, which is the systemic driving force behind these trends as you've described?

There is no doubt that the majority of people do not realise the grave danger of commercialisation, in the same way that most people have no idea how grave and challenging is the reality of the environmental crisis. We can divide humanity into four broad groups with respect to this issue, namely the environmental scientists, the campaigners and activists, the concerned citizens and the remaining masses of everyday people. Among the bulk of humanity who are not engaged with the issue at all, several billion constitute the world's poor in developing countries who are struggling each day to feed themselves and their families, and most of them understandably have no time or inclination to contemplate the scientific reality of global warming, notwithstanding that an increasing number of such people are the unblameable victims of climate-related disasters.

Of the many millions of concerned citizens in affluent countries who are somewhat aware of the climate crisis, a significant proportion of them also have no idea of the real facts or the true scale of the challenge, and they often reduce the issue to personal questions of household recycling or ethical consumerism. But if they knew the true extent of how far humanity has damaged and degraded the planet, or the very real threat of the danger of rising global temperatures, then there would be a lot of tears shed around the world. Just as many scientists are often driven to tears by their knowledge of how alarming is the state of the environment, the whole of humanity would cry for a long period of time if we could suddenly perceive how close we have come to destroying our planetary home.

And yet here is the tragic irony of our situation, as you rightly state, in that the bulk of humanity is even further away from perceiving how commercialisation is at the root of our problems today, which are problems that we are all part of to varying degrees. The reason for this lack of awareness can be understood quite simply, which is to say that the hearts of humanity are not sufficiently awakened or engaged in alleviating the world's crises. Like I have said repeatedly, if enough people really did perceive the true extent of the climate crisis and its inner man-made causes, there would not be just a handful of committed activists among billions of other people. The minds of many concerned citizens may be engaged with this issue, their goodwill may be engaged, but you need to awaken the spiritual heart centre of humanity as a whole—which means you need the active and directed love of millions upon millions of ordinary people. You need the masses standing beside you, hearts aflame. And it's the complacency of the average person of goodwill that is responsible for pushing those stalwart protesters to be only the few.

Is it sufficient to say that our lack of awareness about the true scale of the environmental crisis is simply due to a lack of love or 'heart engagement' with the world's problems, or can we better understand the basic causes of widespread human complacency? If possible, can you also expand on some of the factors that are holding back our awareness of the problem of rampant commercialisation, as well as our awareness of the need for a dramatic and urgent social transformation?

Here again we must contemplate the illusions that sustain our complacent way of life, illusions that tell us 'life goes on', 'be happy', 'enjoy your life', 'you only live once', 'have a good time—you deserve it'. The forces of commercialisation love these illusions, they are like the perfume of business-as-usual, like music to the ears for profit-seeking interests. And these illusions, these fantasies that constantly divert our attention from reality, they have the effect of overwhelming the heart. We could draw an analogy between what a disease like cancer or HIV represents to the human body, and what illusions represent to the human heart. Remember, the heart is like a child—it is very innocent. And the heart doesn't talk or make a sound unless you engage it, because there is no arrogance in the heart, no calculation or self-interest. Only the mind calculates and judges, creating all the divisions of everyday life through our identification with the personality, with the big 'Me'. But the personality is not the heart, and the heart is surely not the personality.

It is often difficult to elucidate these conceptions of love and illusion in relation to the question of social transformation, as we can only perceive and experience the spiritual reality of the heart through inner awareness, and not through books or merely an intellectual understanding. But if you look at this question in the manner I propose, you will see there are two main components that form the prevalent attitudes of complacency and indifference within affluent society—the first being *fear*, and the second being *illusion* which is needed in order to appease that fear. Take the illusion, for example, that we can be independently happy in a divided and dysfunctional society such as ours, and then observe the corresponding fear of loneliness, of not being recognised as a 'success' in life, of the empty finality of death itself. Our constant search for psychological security is the unconscious motivation that underlies our self-seeking desire for happiness, which invariably stems from a basis of fear. In the widest spiritual sense, fear is the baggage that has held back human evolution for thousands of years, like a donkey that is tied to a milling wheel that turns around endlessly in a circle.

And the outcome of these habitual psychological tendencies is what you defined as a lack of heart-engagement with the world's problems, for most people are too embroiled in their crusades for personal happiness and success to realise the truth of 'the great illusion' of commercialisation, or indeed the truth of a world in crisis and confusion. The politicians are confused if not entirely corrupted, and we as citizens are equally confused about the grave reality of the world situation, so the blind are leading the blind. And yet still we vote for new politicians in massive and exorbitant election campaigns, because humanity continues to love its old illusions, especially those that carry with them a sense of continuity or permanence through traditions.

We should view our collective complacency as a grave danger to the world that is on a par with the threat of commercialisation, because don't forget that human complacency was born a long time before capitalist enterprise or economic globalisation came into existence. And when fused with indifference on either an individual or societal basis, we could describe complacency as one of the oldest diseases known to humanity that is even more malefic in its influence than the present day forces of commercialisation. How marvellous that humanity is motivated to discover vaccines for horrific diseases like Zika or Ebola, but we have yet to even search for a cure for our endemic complacency that is assuredly at the root of the world's suffering.

In this light, it is very curious to observe how commercialisation is a grave war within itself, a war against humanity and its spiritual evolution, yet the mechanism of commercialisation is so refined and elusive that many people cannot see the reality of the war, or else they mistake it as merely consumerism. Both consumerism and commercialisation are mutually dependent, and feed upon each other for their continued existence; but it is the *forces* of commercialisation that drive the processes of planetary destruction that we all play a part in today. Just as I consume food, cheap clothes and all the paraphernalia of modern life, I also consume the very idea of becoming individually happy in a divided and ecologically unsustainable society, until I am trapped into a life that revolves around my self-centred search for psychological security through material and emotional attachments. It is a very devious trap that is proposed to us by the forces of commercialisation, one that unconsciously affects every individual by manipulating our subtle or emotional bodies, which is where commercialisation resides as a psychological phenomenon, holding us back through a dense fog of confusion that prevents the mind from functioning clearly with awareness, thus seeing reality as it is. Humanity is so herd-like in its everyday life expression that many of us fail to perceive the pernicious existence of commercialisation at all, while most establishment politicians and financially successful individuals naturally uphold a market forces ideology without questioning its destructive consequences.

Add to this understanding the problem of divisive 'isms' in human consciousness, and we may have a fairly complete picture as to why we are all unwitting participants in the crises of our times. Commercialisation feeds off isms and ideologies as much as it feeds off society's illusions, even if those isms are many thousands of years old. We can also observe the mechanism of how this works in holistic terms, for it is the battle of isms in all their major expressions that perpetuates the dynamic of commercialisation, whether it concerns capitalism, socialism, nationalism, or any other political or religious ism that informs our thinking and actions.

For example, when one group or nation fights an opposing side for nationalistic or ideological reasons, there is always vast sums of money and resources that are involved, such as through the creation and purchasing of more armaments. And the more that weapons of war are bought and sold, the more profits are made through a dysfunctional competition brought about by market forces, which altogether intensifies the power and scope of commercialisation across the world. The rich become even richer, while the poor continue to suffer from the impoverishment and displacement caused by civil wars and regional conflicts. At the same time, competing nations are constantly pushed to increase their control over world resources, while ever prioritising economic growth and corporate profits ahead of all social and environmental concerns.

We could use any number of examples to illustrate these vicious cycles of commercialisation, which is largely sustained by the pre-existing divisions between peoples and nations that originate in our adherence to conflicting isms. There is also a particular phenomenon that is happening across the world today, whereby certain isms are in need of commercialisation in order to propagate their chosen causes and divisive purposes, such as the many religious terrorist groups that employ modern financial strategies to grow in power and scope, often with the covert aid of conniving business and state interests. More generally, we are all somewhat involved in the problem of isms that are increasingly being taken over by the forces of commercialisation, including the problem of racism that has been overshadowed by the process of global economic integration over recent decades. Racism is a form of an ism too, of course, ever rife in modern society regardless of the commercial benefits stemming from cheap labour overseas or less restricted immigration in regions such as the European Union.

Hopefully these brief observations may help you to perceive how commercialisation is about much more than mere business activity or profit-driven pursuits, as it is rather a very ancient expression of those forces of materiality that have subtly re-entered our veins in order to escalate the mayhem and destruction perpetrated by humanity as a whole, from all sides and in all directions. It may seek to destroy the earth and hinder our spiritual growth through the stimulation of mass consumerism; it may subliminally confuse and subjugate the ordinary citizen by propagating the illusions of consumer choice, individual happiness and material comfort; and it may cleverly infiltrate the isms of human consciousness in order to create more wars in the name of deceptive ideologies. It is a dragon of many faces, both perceptible and obscure. And it is constituted by a multitude of different forces, which in their aggregate sustain the greatest of all illusions within the human race—namely, the illusion that humanity can continue to live in such a divided, unjust and self-destructive world order. To observe the psychological mechanisms that commercialisation uses to achieve its aims, it is literally pure evil.

One objection that may occur to some readers is whether progressive activists and engaged citizens are themselves part of the mass confusion, the lack of awareness, and the complacency and indifference that you describe, especially with respect to the environment issue. Are activists not aware of the true extent of this crisis, and often engaged with their hearts in the struggle to bring about a just and sustainable world?

Of the four groups of people I mentioned before in terms of our global awareness about the environmental issue, it's interesting that an informal partnership has now been forged between the environmental scientists and the activists, who are both clearly aware of the daunting scale of the challenges before us, and who both need each other in the face of all that stands in the way of change—by which I mean the whole system of self-interested politics, the corporate capture of governments, the prevalent inertia and public complacency. So if we talk in spiritual terms of love and awareness, then the activists and awakened scientists are certainly engaging the attributes of the heart more than anyone else on this issue, which they express in terms of maturity and responsibility as I previously said.

The many environmental activist groups, in their aggregate, are really the energetic representatives of human goodwill in relation to these issues. But here is the paradox, because they are together fighting a complex system that doesn't allow love to flourish in response to the environmental crisis. In this respect it may be understandable that many activists have become hardened and embittered by the injustice of the system they stand against. After all, they are working on behalf of the entire population, they are representing the common good of all humanity, and yet the population is not supporting them in turn through massive demonstrations and constant engagement with the issue of climate change. So may God bless the world's activists in all their progressive guises, for without them perhaps even the idea of 'environmental justice' would never have come into existence in our awareness.

Please therefore know that it is not my intention to criticise any activists of whatever political or ideological persuasion, but it may be helpful if we look at what is missing from global activism in order to try and intuit more clearly the new ways of engaging the heart. For is it possible that there is a different kind of activism altogether that we can pursue with unprecedented numbers of other people, one that is based on the idea of togetherness, of freedom, of a non-divided approach to human relationships beyond isms and ideology? Unfortunately, what we find is that many activists at the present time adopt a stance of anti-capitalism, including in environmental campaigning and academic circles where the challenge is even framed in terms of 'capitalism *vs* the climate'. This may be understandable to a degree considering the immensity of the forces that oppose us, but it is a grave mistake to believe that we are somehow free from capitalism ourselves, when in fact we are all a constituent part of the system's functioning and processes—a point that I hope was made clear in our earlier discussion.

Hence it is an impossibility to stand against capitalism when that implies we are separate from the problems in our society, and this ill-considered reasoning creates a further 'ism' within itself that is equally a part of the problem. This is a subject that I have written about many times before, suffice to say that there is no love in that understanding—it is just cerebral. The problem isn't capitalism, it isn't neoliberalism, it isn't conservatism or libertarianism—the problem is a lack of love and right human relationship in our world. And the forces of commercialisation are at the centre of this problem today through their pernicious conditioning for people to consume, to become successful, to live materialistically and with indifference to others.

All of our complicated analysis surrounding the various 'isms' will have to dissolve one day, supposing that all peoples and nations can finally learn to live in straightforward peace and harmony with one another. Yet in these overly intellectual times, few activists choose to look at our problems through the inner awareness of an engaged heart, instead choosing to wield the principle of capitalism in such an ill-conceived and fanatical way. It is like the term 'God' which is uttered in so many different contexts, even to say 'God wills it' when we kill other people. Similarly, the term capitalism has been abused and undignified for all kinds of reasons, even by the activists who wrongly assume that capitalism is to blame for society's problems, almost as if a theory or ideology can act independently of human thinking and actions. We can at least be accurate in our condemnation and realise that it is really a war that is going on, a war of violent forces and energies that are more properly termed commercialisation. And what are the driving factors behind these malefic systemic processes? Again we should observe the misguided human intentions and harmful behaviours that we all play a part in, underlying all of which is our sense of separation from each other and the natural world.

Is there a further danger in identifying with a position of anti-capitalism, which is to do with the nature of a system that is so powerful it can easily marginalise or overwhelm those who oppose it? This is another of the main themes in your writings, where you advance the need for a new form of global activism that involves masses of ordinary people uniting behind a concerted call for governments to fully integrate the principle of sharing into global economic policies.[7]

A short answer to this question I believe is common sense, and so we don't need to adopt further complex intellectual theories that only serve to obscure our inner awareness about the simple path towards planetary transformation. A significant proportion of activists who work on environmental and social justice issues appear to stand in opposition to the current system, but just imagine what will happen if masses of ordinary people join them with the same attitude? It is the multinational corporations that control our governments, that have the protection of the police and other authorities, that have the power to shape laws and policies in their favour; so who do we think will win if protesters try to create change through mass antagonism and opposition to the establishment? The more you go against any individual or institution, it is natural that they will take all necessary measures to protect themselves. And in the case of multinational corporations and governments that have the power to create new laws, we are already seeing the foreboding consequences of current trends through global surveillance disclosures.

Of course not all activists adopt the mindset of 'anti' and 'isms', and there's a new phenomenon in the world today which is the pioneering youth who are becoming more engaged, more aware of the need for a wholesale change in public attitudes and governmental policies. They are really the hope of the world, although it's increasingly difficult for them to move in our excessively commercialised societies where everyone is stressed and confused, where everything is being privatised and reduced to a market value. As we know, much of the youth is heavily indebted due to the costs of higher education, many millions more are unemployed and few can afford a decent home. Yet they are predominantly the ones who can show us how to inaugurate the new age for humanity, whatever such a term is supposed to mean. We often hear these terms 'new age' or 'the age of Aquarius' used by spiritual groups, but I prefer to talk in more down-to-earth language about the age of the heart, or the Heart Age, because the heart has always been with us and doesn't need to be given a new label. Besides, if we want to inaugurate a new age, there is no way to do it unless we collectively engage our hearts.

This is one of the reasons why the principle of sharing is of such critical importance to these times, which in a metaphorical sense is like a fountain of milk that all activists from every department of life can drink from, lifting everyone up and unifying us all in our various endeavours. The principle of sharing is made out of common sense, of the will-to-good, of love *per se*. It is what all activists are fighting for in different ways, even if unconsciously, which from the inner or spiritual perspective is about balance and right human relationship. You might say that it is the mother of all attributes of the heart.

PART III

Demonstrating love-in-action

Previously in this interview you explained how environmentalists are beginning to embrace the principle of sharing in their advocacy work, but what does it mean to incorporate the inner perspective on sharing into environmental activism? In other words, how will we know when activists are 'engaging the heart' and demonstrating 'love-in-action'?

It's not the case that environmental activists should stop what they are doing or go in a completely different direction, as it's more a matter of thinking about the inner side of world problems as we continue to pursue our different causes. What this principally means, as I've discussed in the book *Heralding Article 25*, is that it's possible for us to advocate for an irrevocable end to world hunger at the same time as we call for government action to tackle climate change, or whatever other cause we stand for.[8] The basic inner or spiritual significance of the environmental crisis is that it has expanded human consciousness to the planetary level, which means that millions of people are not only thinking of the good of the environment in their own countries or communities, but also the environment of the world as a whole, of the global atmosphere and biosphere that belongs to all of us. It has now almost become fashionable to espouse environmental concerns, whereas a global awareness of the reality of hunger and endemic poverty is seemingly cursed in comparison.

Yet if we can organise big demonstrations for the environment, why can't we organise massive protests for unprecedented government action to feed the hungry and help all those who are suffering from preventable diseases, as we know there are millions of people who needlessly die from poverty-related causes each year? I don't want to repeat the same arguments I've made before, but our awareness of world problems is so fragmented that you can hardly talk to many environmentalists about the crisis of world hunger and poverty, which means that environmentalism is succumbing to another illusion or 'ism' of its own making.

We often hear environmentalists speaking out on behalf of the children of future generations, but what about the children of our present generation who are dying from hunger as a result of publicly-sanctioned government neglect? I submit that those children are in fact dying as a result of the inner CO_2 that pervades this planet, as majorly expressed through our collective complacency, ignorance and indifference. From the most down-to-earth and human perspective, are the children of poor communities not our children too, the ones we should be fighting to care for and protect? The kind of mentality that thinks only of 'my children' or 'my future' in privileged affluent nations is, in itself, the worst pollution that exists in relation to right human relationship. It is really an imbalanced and deluded way to think, far from an understanding of love-in-action or true heart awareness.

I earlier remarked on the danger of politicians putting the environment on the shelf in the event of another global financial crisis, but now that millions of people are focused on the threat of climate change, it's as if we've already put the enduring crisis of global hunger on the shelf. Can we really believe it's a healthy attitude to be concerned about your great-grandchildren's welfare when your family already has more than it needs, compared to thousands of children who are dying each week as a result of living in extreme poverty? However much we clothe these sentiments in knowledgeable opinions about global warming, it is still the same mindset as the complacent person who says 'there has always been hunger, and always will be'. And as long as this mindset continues to persist among a broad swathe of the world population, there will always be those who say in response to the climate crisis: 'Well who cares at the end of the day as we're all doomed in one way or another'.

There are also many prominent climate activists and campaigners who never mention the crisis of global hunger and poverty, as if they are more concerned about decreasing the world's average temperatures than in feeding the world's dying poor. Hence they effectively stand for the common good of those who live in modern affluent society, but not for the common good of the world as a whole, of the one Humanity. That is why, as far as I am concerned, the foremost spiritual authorities on this issue are Willy Brandt and Pope Francis respectively, as reflected in the Brandt Report and the recent encyclical, *Laudato si'*.

In the third chapter of *Heralding Article 25*, you postulate that "we can never tackle climate change or the environmental crisis without also remedying the injustice of poverty amidst plenty, which is where the solution to our manifold ecological problems initially begins". Can you summarise your line of reasoning?

I would urge readers to contemplate for themselves this line of reasoning, but the essential message is that constant demonstrations are the only way to reorient the critical world situation through a massive expression of unity among millions of people in all countries. *Article 25* of the *Universal Declaration of Human Rights*—which postulates that every man, woman and child should have access to the basic necessities of life—is truly a great guide for the masses of ordinary people towards justice and freedom, considering the prodigious changes that implementing *Article 25* would augur in global political and economic affairs.

This is particularly with respect to the United Nations which we should regard as an organisation that belongs to the people of the world, not to corporate interests or the major powers represented in the Security Council. In this way, huge and constant protests around the world that revolve around the human rights of *Article 25* could eventually bring the Brandt Report back to the negotiating table, or at least its high-level case for an emergency programme of redistribution to end hunger and extreme poverty as a leading international priority. That is, in essence, the gateway to further structural changes to the world economy that are necessary for tackling our longer term security and environmental crises.

Anyone can understand and follow the brief rationale I posit in the book, central to which is an understanding of the problem of commercialisation as we've discussed at some length in this interview. Without ceaseless protest activities among millions if not billions of people, on and on for many months and even years, then the forces of commercialisation that dominate our political and economic institutions will continue to intensify world trends towards further catastrophe and turmoil.

In the present context, it is impossible that nations can implement an ambitious agenda for limiting global carbon emissions, for permanently ending hunger, even for sharing the burden of dealing with the refugee crisis. But once governments commit to sharing global resources to end the moral outrage of life-threatening poverty in a short number of years, there are many other issues that nations will have to cooperate and share in resolving—including issues that pertain to tackling climate change and healing the environment. In the book, I use simple deductive reasoning to explain this indisputable logic, which includes the vexed issue of how to stabilise and eventually reverse the world's population levels by permanently securing every person's basic socioeconomic rights.

However, this is not another theory of change that I'm trying to promote, as these are down-to-earth and human questions that do not require us to move within any theory or 'ism' to understand them. The solution to world problems as I propose is remarkably straightforward in its conceptualisation, but the deeper questions concern what will happen if millions of people protest continually with a single aim and unified purpose, which are questions that take us back to the 'inner' line of enquiry about our lack of love and heart awareness. If we can foresee endless demonstrations that focus on empowering the United Nations to oversee the implementation of *Article 25* in all countries, then it will have the effect of bringing millions of people to more awareness, to see more clearly the reality of the critical world situation.

We have to come to a collective realisation that all the problems we are witnessing today, all of the misery, division and hatred, is due to the fact that we have failed to share the world's resources since the creation of the United Nations, and we continue to go headlong in the wrong direction. It is not *Article 25* that is the solution to world problems in this sense, but the awakening of the human heart on a planetary scale through awareness, togetherness and love, and through the realisation that we have to implement the principle of sharing into world affairs before it is too late.

If enough people arrive at this joyous and hopeful realisation, then we can envision how our collective awareness will soon be expanded to realise the connections between the world's interlocking crises. Once the majority of humanity is wholeheartedly concentrated on resolving the problem of hunger and poverty, we can more readily perceive the injustice that underlies our environmental problems in terms of man's exploitation of man, in terms of the legitimated theft and destruction brought about by multinational corporations. In time, millions of ordinary people will be led to question the obsession in our societies with material goods and wasteful consumption, and will instead seek to live more simply without constantly striving to become rich and successful.

It will also have a tremendous effect on our educational systems as well as our cultural and social institutions; indeed, the effects of securing everyone's basic needs will eventually uplift and benefit all aspects of human endeavour. I am only briefly summarising what is an immense and extraordinary transformation that will happen in our societies if we finally implement *Article 25* within the course of a generation, which will no doubt be a difficult and long process even if a majority of the world supports this formidable cause.

You just alluded to a significant issue that you've only touched on so far, which concerns how we may actually realise less resource-intensive lifestyles in affluent countries and more equitable living standards between and within all countries. Although you briefly addressed this issue from the exoteric or 'outer' perspective in the first part of this interview, there are many questions remaining about how we can bring about the requisite changes in our attitudes and behaviours within modern consumerist societies. Is this also a subject that can only be truly understood from the inner or psychological perspective, in terms of a shift in human consciousness or an expansion of our collective awareness?

It's true that few progressive thinkers examine the politics of this question from a more holistic, spiritual or psychological perspective, which is the only place from where we can perceive a real answer. The first thing to understand is that in order for a population to live more simply and equitably within the biophysical limitations of the earth, you have to decrease the incidence of stress and depression throughout the world as a whole. A huge proportion of humanity is suffering from stress due to the strain of living in a dysfunctional society with its endless divisions and conflicts, and the result of all that stress is fear in every respect—fear of the other, fear of change, fear of life, the fear of *being* itself. And you cannot expect a person who is suffering from stress, depression and fear to live more simply, when this also requires us to be more inwardly free and detached.

To carefully observe someone who is suffering from stress and depression, their psychological tendency is to try and seek a sense of inner security through attachments, whether they be material or emotional in their nature—which is clearly the opposite of pursuing a free and simple life with awareness. One of the cures of the prevalent harried way of life is through constant loving attention in all areas of one's daily activity—but here again is our predicament, for how can you counterbalance the stress of our society when our entire world lacks love, lacks spirituality, lacks even the knowledge of inner freedom? Hence the absurdity of our situation, for we are denied the very possibility of living simply when that is the only lasting solution to our common problems. At the Krishnamurti schools, for example, many teachers have lamented about how their pupils have learnt to pursue a simple, intelligent life with right values and right livelihood, until they end their education and enter a dysfunctional society that knows almost nothing of what it means to live intelligently, lovingly and with awareness.

Also as a result of the fear of being, it's notable that many people try to become more psychologically secure by fighting back against society with all of its inequities, hence the many 'isms' and ideologies are constantly reinvigorated and looked towards for refuge. This is a very important subject that we would all do well to reflect upon for ourselves in the *grosso modo*. Today, it is not only the fear of being that is preventing many people from changing inwardly, but also the fear of losing their isms, or the familiar ideas which form an integral part of their identity—mainly of a religious, political or cultural persuasion. Following this line of inner enquiry may also help to explain why the mass confusion of our times is so grave and debilitating, for when confusion becomes a worldwide epidemic it significantly decreases the creativity within mankind, which further prevents the necessary transformational changes from taking place within society.

Another important aspect of this issue to contemplate is that we cannot simplify our lifestyles *en masse* without experiencing the joy of living, and yet there can be no true joy in a dysfunctional society that is so unequal and unjust, and so confused and corrupted by ignorant politicians. There are two kinds of joyful living that we can discern and broadly differentiate, the first one stemming from the superficial pleasures and illusions of commercialisation, which in its aggregate is the greatest of all illusions as we've now discussed. But the other joy of living, the true and spiritual *joie de vivre*, is unknown to almost everyone today, for it can only become apparent and reveal its nature once the principle of sharing is fully integrated into global economic and political affairs.

It's as if these two opposing modes of living each cancel the other one out, and today the illusory joy of living that is proposed by the forces of commercialisation is the one that humanity has been unconsciously pushed to adopt. Hence the joy of living that is proffered by the heart via compassion, via awareness, via the joyful recognition that we are all one, that we are all equal and essentially the same in our togetherness and love—that is a mode of living that is eclipsed by mass consumerism and largely unfelt in our daily goings-on. We only experience the faintest glimmerings of that joyful way of being in our various expressions of sharing and solidarity on a community level, or in the makeshift encampments of peaceful mass protests such as Occupy Wall Street or Tahrir Square in 2011.

Does this mean that a simpler and more joyful way of life within mainstream affluent society can only come about through the implementation of *Article 25* across the world, which is a subject that you introduce in the book *Heralding Article 25* but don't expand upon in detail?

If you can follow what I'm saying not only intellectually but also through the awareness of the heart, then you can see that to live simply is quite something in this excessively materialistic world, and it is the implementation of *Article 25* that will guide humanity in how to live more simply within the means of our shared planet. Does it not occur to you that we can never live more simply and equitably in our respective standards of living worldwide, so long as millions of people are needlessly dying each year as a result of hunger and other poverty-related causes?

There are many extremely wealthy people who are tired of being rich in their segregated surroundings, just as there are many millions of poor people who are tired of their daily struggle to survive, and it is the prospect of fully guaranteeing *Article 25* that can bring all of them together and lessen the tensions that are caused by the huge discrepancies in living standards worldwide. In psychological and spiritual terms, *Article 25* represents a phenomenal basis of dignity and freedom that holds the potential to lessen the incidence of greed, dishonesty and ambition in our attitudes to life and human relationships, which is the precursor to releasing the love and awareness that is needed to simplify our various modes of social organisation.

After all, what is the basic cause of climate change if not the abuses caused by the pursuit of profit and power, for it is impossible to make profits on a colossal scale without abusing the environment and exploiting other people. We have observed how the governments and big corporations are complicit in the maintenance of this destructive economic order, just as we are complicit in sustaining the whole system through our mass patterns of unsustainable consumption. But it seems that almost everyone wants to consume more and increase their relative standard of living, including the millions of poor people in developing countries who aspire to enjoy an overconsuming lifestyle like those who live in Western Europe or North America, which as we know is disastrously unsustainable in terms of 'One Planet Living'.

The only way to live simply in this society is to go off-the-grid or retreat to an isolated community, but that is no solution to the global problem of climate change that many scientists believe has already passed beyond certain irreversible tipping points. Whatever benefits we may derive from living more frugally in self-sustaining communities, it is really just an escape that can only be short lived because, sooner or later, the world's problems are going to come and find us wherever we live.

This points to another of the spiritual significances of climate change, which is to teach humanity that all of our problems today are global problems that demand the world as a whole to join together in resolving them. It's as if even the weather is asking us to stop dividing ourselves from the spiritual unity of mankind, meaning that we can never bring about a simpler and more joyful mode of living until enough people engage their hearts with the interlinking issues of ending poverty and saving the environment.

What you seem to be saying is that it's impossible to realise simpler living standards worldwide until the implementation of *Article 25* is first prioritised in world affairs, because this is what will release both the 'inner' awareness of the need for humanity to change, as well as the 'outer' structural changes in society and the global economic system that need to be based on the principle of sharing?

Whichever way we look at this issue, we cannot move forwards in our thinking and visionary proposals for a new society until we understand the necessity of implementing *Article 25* as an inviolable set of laws within all nations, as this is indeed the gateway to both the 'inner' changes within humanity and the 'outer' changes within society that can ultimately bring about a sustainable economy that operates within planetary limits. What else do we think will lead us to the collective awareness that is needed to sustain new laws and new economic practices that prevent the overexploitation of scarce natural resources? When everyone has their basic needs secured, when resources are distributed more equitably across the world, then perhaps we can foresee the introduction of a new international governance framework that will enable us to protect the environment and live free from injustice, thus realising the joy of living more simply 'so that others may simply live'.

We are really talking here about the fusion of capitalism and socialism through entirely different economic arrangements on a global level, whereby the implementation of *Article 25* will determine the future trade arrangements between nations so that if a country has a surplus of food, for example, it will not seek to export that food for profit or leave it to rot in vast storehouses while millions of people are hungry elsewhere in the world. As an alternative economic paradigm, any nation with a surfeit of produce in excess of its domestic needs may donate it to some form of global pool that is managed by an institution formed specifically for that purpose, such as a new specialist agency established within the United Nations, from where it can be redistributed or exchanged with other countries that produce much less of a certain resource.

We might think of this as the true structural adjustment, one that can enable the community of nations to produce a sufficiency of goods and produce that everyone has the right to enjoy, without exhausting the limited supply of the earth's non-renewable resources. But how will this be possible to achieve with the continuing explosive growth of the world population, with the growing extremes in living standards among the richest and poorest citizens, and with the increasing overconsumption of natural resources that is already far in excess of the planet's biocapacity?

We could endlessly debate the details of a simplified economic system that is founded upon the principle of sharing, but the fact is that no structural reforms or new laws can be introduced along these lines while humanity is so conditioned by the forces of commercialisation, behind which lies the all-embracing pursuit of profit and competitive self-interest. What is the psychological factor that holds back the environmental commons from flourishing in the consciousness of humanity, if not the pursuit of profit? And what is the inner impediment to establishing cold fusion technology as the basis of free energy within our societies, if not the pursuit of profit? By stretching our imagination far into the future with awareness of the inner or spiritual reality of life, we may come to the conclusion that the only way out of this age-old problem is to structure the practice of bartering into global economic affairs within the law of demand and supply. The pursuit of profit is harmful and stressful to both man and nature, as we know, whereas the practice of barter is based on the opposite propensities towards harmlessness, simplicity, and right human relationship.

There is a great deal more to be said on this topic, but the point I'd like to emphasise is that such a new mode of economic exchange will be impossible to sustain without the concurrent introduction of a new education based on Self-knowledge, which is a vast subject in itself that requires an entirely different way of looking at our relationships to each other and the natural environment. We are obviously not talking about going back to more primitive lifestyles, and there will always be a role for innovation and technology in the evolution of human societies, while there is nothing that can stop the innate curiosity of man to learn about his place in the universe. The need for simpler standards of living also concerns much more than the amount and type of resources we consume, for there is an art to living simply that implies an awareness of the inner Self in our relationship to the outer world. Please keep in mind here that a simpler and more joyful, more heart-engaged way of life cannot be imposed on people by any external authority, and it must be cultivated from within each and every person through an inner awareness of one's intention to live harmlessly and with respect for all other living beings.

Please expand further on these last sentiments if you will, particularly the idea that there is an art to living simply. Are you talking about the same 'Art of Living' that John Stuart Mill envisaged when he contemplated the end purposes of industrial progress in the mid-18th century? Or is there a different meaning to living more simply as an art form, which presumably must be understood in a spiritual and not a literal sense?

To live simply is verily an art, in the same way that an accomplished architect can design a building that is basic in its construction and frugal in its use of resources, yet beautiful in its outer appearance. Or like the artist who knows how to put different colours in their right place so that a painting is visually appealing and meaningful, we too are trying to put this society in its right place so that everyone has what they need for a dignified, creative and fulfilling life that is completely harmless and oriented towards the service of others. If we believe that there is something more to our existence beyond the stress and divisions of the modern world, if we can sense that a different way of living is possible in which everyone has the freedom and opportunity to explore their inborn creative potential, then we can intuit the truth that the whole of human evolution is like a scientific art where we all play our role as co-creators of the One Life. We are all artists of our own creation, from the inner to the outer. And there can be no such thing as 'living simply' without the Art of Living, or what is more accurately expressed in spiritual terms as the Art of Being.[9]

We can also observe that at this time of planetary chaos and transition when it is nigh impossible to live simply within our day-to-day life expression, then the first step for humanity is what we might call the art of right thinking. This is, after all, what can make us into an artist of our lives—by firstly becoming aware of the spiritual reality of life, and of the need for a complete change in our attitudes, behaviours and intentions as a family of nations. If I identify with the idea of being a successful and rich 'entrepreneur', for example, then I am more likely to care for opportunities to make profits and amass personal wealth, than I am to care for the viability of public parks and protected nature reserves, let alone the stability of the Earth's climate system. The art of right thinking therefore concerns my originating intentions, which must invariably be oriented towards the greatest good of humanity and the planet as a whole.

There are already many artists in the world, in this sense, who are concerned with changing how humanity conducts its national and global affairs so that we may better realise the collective wellbeing of all people as we transition towards ecologically resilient societies. If we can imagine the existence of progressive civil society groups and humanitarian agencies as representing certain noble ideas about right human relations, then the art of right thinking concerns the rendering of these ideas into actuality through the mass engagement of ordinary citizens. An end to poverty, a sustainable economy, a peaceful world and a healthy environment; all of these ideas will never come to fruition until the whole of society unites behind them, as we have thoroughly acknowledged throughout this discussion. And it is the embracing of these ideas by humanity in its entirety that will expand our consciousness as a race, making all our lives more joyful and guiding human intelligence to unfold at a much faster pace in alignment with our spiritual evolution.

Thus the greatest disease on this earth is *separation* in both its inner and outer forms, and the art of right thinking concerns the very beginning stage of our cooperative planetary transformation. I do not believe we need to be too preoccupied with the art of living more simply at this time, as it is much more important to become aware of the illusions that are leading us astray from the simple path of right human relations. If we want to further contemplate these prognostic spiritual questions of what it may be like to live in a more enlightened and heart-engaged society, I suggest we start by becoming more cognisant of how the increasing trends of commercialisation are taking us in the opposite direction. To extend the analogy of the visual artist, it is commercialisation that is wrongly educating us, conditioning us and preventing us from putting all the colours in their right place, until our misapplied human intelligence is leading us towards our imminent self-destruction, as I have said many times.

All this leads to the question of what can awaken the love and awareness of humanity to a sufficient degree, thus initiating the process of world transformation as you have proposed through massive civic engagement. Is it possible to predict the moment or spark that may signify the onset of these inner and outer changes? We still appear to be far away from your vision of millions upon millions of people uniting through constant demonstrations for a fairer sharing of the world's resources, and this may leave many readers with a continuing sense of hopelessness and confusion about the world's problems.

If we examine only the outer problems of humanity then we are faced with such complexity, and so many factors and variables, that we cannot possibly fathom a clear answer for how to resolve the world's problems. This is why it is of benefit to ponder the inner or spiritual side of human evolution which will lead us to deeper, simpler and more transformational insights about how the world needs to change. For thousands of years, many wise teachers and prophetic thinkers have implored man to look inwards and perceive the need for change, and yet still we have failed to learn the most basic lesson of how to treat each other with kindness and affection, how to share the produce of the earth without self-interest, competition and greed.

Do we think humanity has learned anything about right relationship after all these centuries of technological and social advancement? Nations have tried and abandoned many 'isms' that informed our social behaviours, we have endured great wars and many centuries of colonialism and imperialism, and yet still we continue in the old ways of aggressively competing for power on the world stage, with all the resulting divisions and inequalities.

So what do you think it will take for all the people of the world to realise that we are one Humanity, inherently equal and interdependent as a fact in nature? From the inner perspective, it isn't the economic and political safeguarding of *Article 25* that holds the solution to our problems, because all of the world's problems are interconnected and from the same anthropological origins, which means there is only one problem and one solution in the end. The problem itself *is* the solution, which in the most basic spiritual and psychological sense can only be expressed, once again, as a lack of love. Let us say that we can never find a vaccine for this ancient disease which afflicts humanity, for only the human heart holds the antibody.

It appears that we have still to learn how to live with one another without creating new forms of mind conditioning, without hiding behind the masks of our religious and political 'isms', without reproducing the prejudices and hatreds that are buried in our nation's history. All we can finally say is that right human relationship is the key to resolving our social, economic and environmental crises in all their dimensions, which in the context of our dysfunctional societies means that we need a copious amount of love and awareness to be released throughout the world.

But as long as we continue to vote for the same politicians that uphold the old ways of national pride and individual prosperity, as long as protest actions and demonstrations remain confined in numbers to the relatively few, then there are only two prospects for a rapid expansion of human awareness: either divine intervention, or a total collapse of the international economic system. As I wrote at the end of *Heralding Article 25*, it seems that a great catastrophe is necessary to bring us to a sense of reality, to completely disrupt our current way of life and all that lies behind it in psychological terms—the tsunami of personal ambition; the drive of millions of people to be rich and successful; the individualistic pursuit of happiness and security. Another global financial crisis that is severer and more protracted than anything experienced since the Great Depression will have a profound effect on how we think and organise our societies. It will demonstrate, in a symbolic sense, how humanity has reached a point that it cannot go beyond, and there is no way forwards without a complete transformation in our inner attitudes and behaviours which sustain a social order that is predicated on the wrong premises, the wrong principles.

What else can stop the juggernaut of mass consumerism and shake enough people out of their complacency? But to reach this point when it is necessary for humanity to experience an unmitigated economic crisis is really a pitiful indication of the depths to which we have fallen at the onset of the twenty-first century, until there is no trust among the nations, no love expressed in our governmental institutions, and no vision among our political leaders of a genuinely cooperative way forwards. We should ask ourselves why we even need massive and constant demonstrations to awaken sanity in global affairs, when it was always so simple to change how we live together and act as a human race.

You have succinctly mentioned in various ways the spiritual significance and symbolism of climate change; is there anything further you would like to say about the inner causes and implications of the environmental crisis?

This is another important question that we should reflect upon further for ourselves, because if everything is spiritual in this world and man is Life, as I have asserted, then it means that everything we do has an effect on the environment around us, from the individual actor to the entire human collective. The chaos within the environment reflects the chaos within ourselves, and hence the crisis of global warming is directly the consequence of our amalgamated thoughts and actions. What you sow, so shall you reap, as it is written in the bible. How simple can it be? But as a civilisation we have yet to realise and accept that nothing belongs to anyone, and hence the resources of this earth are the responsibility of us all to fairly distribute, steward and protect. Once we learn and demonstrate that essential first step into our true humanity, our shared divinity, then we may see how nature returns to its balance in a mysterious and even miraculous way. Man and nature are eternally one in spiritual and evolutionary terms, so if division is brought about by man's hubris and ignorance over many centuries, then the whole of creation goes out of balance, a balance that was naturally generated and sustained before man came into existence.

It was in this light that I remarked upon some of the great symbolism represented by anthropogenic climate change, in that it reflects how humanity is in crisis on every level—not only economically and politically, but also psychologically, socially and spiritually. So we are not only witnessing a political crisis that affects the operation of our governance institutions, but also an 'isms' crisis that calls into question our entire approach to human thinking and action. Similarly, it is not only a renewed global financial crisis we may soon experience that calls for new modes of economic organisation, but also a culminating psycho-social-spiritual crisis that compels us to reassess the very nature and purpose of our being.

All the old isms are collapsing or crystallising, just as all the old forms of our economic and political systems are slowly melting or breaking down. Hence all the confusion among conflicting and polarised perspectives, the left wing and the right wing, the reactionary conservatives and the staunch progressives. As humanity undergoes a painful and chaotic transition to a saner world, it appears as if we are living in an era of mass confusion. And depending on our response to the manifold crises that mark these difficult times, it is possible to predict that this period of endemic confusion is bound to intensify for many years to come.

Is this the definitive meaning of climate change from an esoteric point of view—that the only way out of this era of confusion and crisis is through the implementation of the principle of sharing in world affairs?

As I have said from the beginning of this interview on the inner causes of our environmental problems, what it essentially means is that humanity is being presented with an ultimate and unavoidable choice. Either we continue to go in the old selfish and competitive ways of the past that will eventually lead to self-annihilation, or we embrace the new forces and energies that are dividing humanity into two opposing sides, thus leading every right-thinking person of goodwill towards a progressively more obvious realisation—that we have to share the world's resources as our last remaining hope.

Both sides in this epic struggle are confused and frustrated, not only the activists in all departments of life who are fighting against the self-interested governments and corporations. Even the proponents of commercialisation don't know what to do with these new energies that they can sense and somehow perceive, however dimly or unconsciously, and from which their profit-making and personal ambition is the only source of relief. We are all in the same boat when it comes to our mass confusion regardless of what isms we cling onto of the left or right, because humanity is now coming of age—which in the most symbolic sense of all means that we are finally ready to accept the truth of our innate divinity and oneness, beyond any religious or theoretical interpretation of what these terms are supposed to mean. It's as if the weather is not only asking us to stop dividing ourselves from our spiritual unity and interconnectedness, but it's also imploring us to look at ourselves and perceive how we contribute to the inner causes of the world's crises, for this is our final opportunity to collectively change from within.

To recapitulate, you've explained that global transformation cannot come about through the ideas or advocacy work of civil society groups alone, however sensible and commendable their proposals: it can only emerge in the wake of a shift in consciousness among a majority of the world's people. And this new consciousness and heartfelt awareness will need to be expressed, first and foremost, through massive popular protests that embrace the moral, spiritual and practical urgency of guaranteeing *Article 25* for all people in all countries. That is the gateway to empowering the United Nations and ushering in a new era of international economic sharing and intergovernmental cooperation, as you've outlined with reference to your other writings. Thus somewhat paradoxically from the inner or spiritual perspective, it means there can be no enduring solution to the environmental crisis unless we also embrace the need for an urgent end to human material deprivation. Is that a fair summation of your thinking, or can you clarify or add anything further?

The whole thought-form about the global ecological crisis is becoming more and more crystallised now, because its spiritual counterpart is hardly mentioned in most activist circles and popular debates. I have argued elsewhere that climate change is the only teacher we currently have that is slowly uniting humanity together on a single issue, at least in a global and symbolic sense. But if we are to realise and truly demonstrate that humanity is one interdependent family, we cannot continue to ignore *Article 25* as our foremost and dual priority in the fight to save our planet. It should be increasingly obvious to any engaged citizen or politician that the rapid safeguarding of *Article 25*—which inherently calls for a massive redistribution of global resources and a wholesale restructuring of the international economic architecture—is absolutely linked to the solutions for climate change. For if there was even a modicum of social justice in this world through a fairer sharing of resources, if no person on earth continues to die of hunger or other poverty-related causes, then the weather would not be in such an imbalanced and chaotic state as today.

As I have emphasised, we can understand this logically and deductively in *outer* or policy-related terms, as per some of the initial reflections in this interview on civil society proposals for how governments should address the climate crisis through an equitable effort-sharing framework. But for a real and heartfelt understanding of the world's environmental problems we are called upon to engage with our *inner* awareness, our intuition, our compassion and our common sense, that is if we want to perceive for ourselves the significance of the great link between the hidden elements of nature and the aggregated thoughts and actions of humanity as a whole. Perhaps there is no other way to resolve the world's problems unless a major proportion of humanity first of all succeeds in engaging with the heart by heralding *Article 25*, which may be the only route to escaping the systemic impasse brought about by the current growth-based economic paradigm, one that we are all complicit in sustaining by our adherence to the forces of materiality and commercialisation.

So I believe it is of the utmost importance that we contemplate and heed the *inner* significance of implementing the principle of sharing into world affairs, which does indeed present a certain paradoxical difficulty as this requires us to examine these issues through the awareness of the heart, when the heart *per se* cannot be intellectualised. It therefore behoves us to reflect again with renewed attention, inwardly and quietly by ourselves, on how we need to integrate the principle of sharing into our global economic arrangements so that we can gain trust between the peoples of different nations, which will enable us to release the attributes of the heart much easier and faster in the time ahead.

Naturally, the forces of commercialisation stand opposed to the manifestation of love and wisdom in this world. But when the attributes of the heart are released on an unprecedented scale, then a new phenomenon will arise in our societies whereby the creativity of individuals will be tremendously enhanced; the tensions and stress in the world will dramatically decrease; the joy of living will become a palpable and universal reality; even the healing of diseases will take a new and more rapid course of advancement, due to the lessening of depression and psychological suffering in all its forms.

In short, the implementation of the principle of sharing among nations is our greatest hope and augur of a better world, not least with respect to the environmental crisis, as it will guide humanity to go in a different direction from the inner to the outer, from the spiritual heart centre to the consecrated mind, from our interior awareness of Life to the exterior balance of the world around us. These are some of the imperative questions I will explore in more detail in future *Studies on the Principle of Sharing*.

As a final question, a lot of people believe that humanity doesn't have the capacity or inclination to change on the scale required, and we are therefore heading towards a dystopian future over the coming century. Do you personally have hope that we can make it through the great transition that shortly lies ahead?

There is every reason for hope. Many people lost hope during the Second World War, and yet the allied powers won over the forces of darkness as manifested in the totalitarian regimes of Germany, Italy and Japan. Now we have to win over another dark and formidable power—and that is the forces of commercialisation, as discussed throughout this interview, which have increasingly gripped every society through an extreme market-oriented ideology that is leading our civilisation to destruction. It may seem as if we need divine intervention in order to transcend the coercive influence of these forces in every aspect of our lives, but we have defeated those materialistic forces once before, and we can defeat them twice. If need be, we can defeat them 10 times. Because if there are forces of darkness on this earth, there must surely be forces of Light. And we shall stand by those forces of Light once again, for the purpose of our existence is to spiritually evolve with dignity, equality and in freedom.

So my response to those people who doubt we will make it is: there is hope, great hope, that humanity can overcome this epochal crisis of our civilisation. We have not yet witnessed the power of sharing as a global phenomenon when it is expressed among millions upon millions of people through the heart with its attributes of love. And when the principle of sharing is genuinely incorporated into the policies of the world's governments, then dramatic changes will begin to take place in our political and economic structures with a rapidity that may startle all of us. Thus we may finally understand that the greatest curse for nature, throughout all these centuries of exploitation and destruction, is that humanity has always refused to come together and act on behalf of the common good of all. For life is One, and man is Life itself; there is nothing more to say.

ENDNOTES

1 Willy Brandt, *North-South: A Program for Survival* (The Brandt Report), MIT Press, 1980.

2 "At the deal's heart must be a settlement between the rich world and the developing world covering how the burden of fighting climate change will be divided—and how we will share a newly precious resource: the trillion or so tonnes of carbon that we can emit before the mercury rises to dangerous levels." See the full editorial: 'Copenhagen climate change conference: Fourteen days to seal history's judgment on this generation', *The Guardian*, 7th December 2009.

3 Encylical letter, *Laudato si': On care for our common home*, Libreria Editrice Vaticana, May 2015.

4 'A discourse on isms and the principle of sharing', Share The World's Resources, July 2014, <www.sharing.org/discourse-isms>

5 Mohammed Sofiane Mesbahi, *Heralding Article 25: A people's strategy for world transformation*, Troubador Publishing Ltd, 2016. See Part V: Education for a New Earth.

6 cf. 'Christmas, the system and I', Share The World's Resources, December 2013, <www.sharing.org/christmas>

7 See for example: 'Rise up America, rise up! A letter to an American activist', Share The World's Resources, October 2014 <www.sharing.org/rise-up-america>; 'Uniting the people of the world', Share The World's Resources, May 2014, <www.sharing.org/uniting-the-people>; *Heralding Article 25*, op cit.

8 *Heralding Article 25*, op cit; see Part III: The environment question.

9 This subject was also discussed in 'The true sharing economy: Inaugurating an Age of the Heart', Share The World's Resources, November 2016, <www.sharing.org/true-sharing-economy>

10 *Heralding Article 25*, op cit, p. 65.

ABOUT THE AUTHOR

Mohammed Sofiane Mesbahi is the founder of Share The World's Resources (STWR), a civil society organisation based in London, UK, with consultative status at the Economic and Social Council of the United Nations. STWR is a not-for-profit organisation registered in England, no. 4854864.

The interview was conducted by Adam W. Parsons, STWR's editor. For more information about STWR, please visit www.sharing.org